Worship, Warfare & Intercession

Before the Throne of God

By Robert Misst

Copyright © 2011 by Robert Misst

Worship, Warfare & Intercession
Before the Throne of God
by Robert Misst

Printed in the United States of America

ISBN 9781613793206

All rights reserved solely by the author. The author guarantees all contents are original and do not infringe upon the legal rights of any other person or work. No part of this book may be reproduced in any form without the permission of the author. The views expressed in this book are not necessarily those of the publisher.

Unless otherwise indicated, Bible quotations are taken from The New American Standard Bible®. Copyright © 1960, 1962, 1963, 1968, 1971, 1972, 1973, 1975, 1977, 1995 by The Lockman Foundation. Used by permission. (www.Lockman.org); and The HOLY BIBLE, NEW INTERNATIONAL VERSION® (NIV). Copyright © 1973, 1978, 1984 by International Bible Society. Used by permission of Zondervan.

Scriptures in **bold** print is the author's emphasis

www.xulonpress.com

Acknowledgments

My thanks go out to so many people. First, to my wife, Marinelle, and my son Yohann, whose patience, prayers and love have carried me through this time of writing my first book, taking on my house chores to release more time for me to write. Thank you, my dear family, for understanding me, and all the frustrations of getting it right first time around.

Jill Shannon, my sister in Yeshua our Savior and Lord, is the co-creator of this project. Her help, passion for worship, and deep insight into the life of the Spirit, produced the most awesome CD "Sounds of Heaven." Her CD supports the "worship in heaven" chapters of this book. Jill has also written the Foreword, as well as Chapter Four of Part I of this book. Chapter Four keeps in mind young songwriters, and encourages them to write songs of worship. Jill has been the most loving, patient, and skillful editor of this book. Her extensive studies in the Word of God, ministry experience and her deep love for the Lord have shaped this book to share its message in a way that will be a blessing to its reader. I am deeply indebted to her for her time and resources in the midst of her extremely busy schedule. I pray that this book and Jill's CD, *"Sounds of Heaven,"* will be a blessing to many in the body of Christ. Jill, it has been an awesome experience working with you on this project. You are a blessing not just to me, but also to the entire Body of Christ worldwide. Thank you for being so kind and patient, and going through many sacrifices to encourage me all through this project. I cannot thank you enough for your efforts to see this project give its very best for our Lord Jesus, all glory to Him alone.

My thanks go out to so many people in the body of Christ. To those who trust me, as we share our spiritual experiences and journey. I cannot fail to mention prophets and intercessors like Warren Lyons of Christchurch, New Zealand, the founder of the National Days of Prayer. To my intercessor family, Bill Hotter and Bryan Chivers of the Canterbury Regional Intercessory network, New Zealand. My wife and I share our life with many different intercessory groups. My thanks to members of: the "Auckland Regional Intercessory Network," House of Prayer Emmanuel – Intercessors of Auckland (HOPE, to mention a few: Kate McGiven, Mary Power, Ajit and Patricia Pinto, Michael and Philomena Lobo, and to all who are part of this intercessory ministry, thank you for your love and support). To many members of the Lamb of God Ecumenical Christian community, whose labor of love has helped us when we need help, and many others in this ministry of prayer, who I may have left out. Also, to all those lovely musicians that have been patient with me, as I have tried to explain to them the sound and songs of heaven, when I am so ignorant of musical terms, and have difficulty singing in the right key. I am sure that Jill's worship music will communicate these truths to you in a way that exemplifies my written words.

It has been a long journey for me in the ministry of prayer before the throne of God. To be honest, often, I tried quitting. Nevertheless, God is so patient, gentle, and understanding of my fallen human nature, that He has drawn me back. Sometimes I do feel like a dropout from the school of ministry. In the school of His Holy Spirit, where I doubt I will graduate to the top of the class, I have learned to be patient, and see life through the eyes of eternity. In His time, yes, He will make everything beautiful in His time. So if you run into spots in this book which may be a bit unclear, remember we are dealing with the spirit realm, and it could leave one in that state. I understand that there are some ideas expressed in this book, which not every reader will agree with. I can certainly respect that. Everyone has a right to his or her opinions. However, I am sharing these revelations and teachings as accurately as possible, because

I have received them from the Lord, and He desired that I should write this book.

Blessed are the pure in spirit, for they shall see God. The Lord knows our hearts, and He understands how each reader approaches the ideas in this book. What He wants is a pure heart. I have endeavored to write it with this in mind – keeping a pure heart before Him. My prayer for you, dear reader, is that the Lord will richly bless you as you minster to Him in worship, and to the world through humble intercession, until He returns. He is coming soon; there can be no doubt about that. The writing is on the wall!

Endorsements

This book is not for the fainthearted! It is an urgent call to the church in every nation to prepare for, and hasten, the return of Jesus Christ as King of the Nations. The book is not designed to bless and make you feel good, but to challenge and change the reader! It could make most pastors nervous as Bob rightly observes that to follow the teaching would result in a 'new overhaul' of how we do church. In 2010 Bob experienced in a vision the worship of Heaven around the Throne of God. He heard and saw remarkable things in the Spirit that lined up with the book of Revelation. He saw that in heaven there were seven key aspects to worship as illustrated in the book of Revelation. Bob's burden is to discover Heaven's agenda and help the churches to make it their agenda. To quote:

"When the church patterns her worship, warfare and intercession after heaven, she joins heaven in the worship that initiates the restoration and renewal of the earth."

This is serious stuff and demands a serious response from the church in the context of God currently shaking the Heavens and the Earth, and bringing chaos to the Nations prior to the soon return of Jesus Christ. There is rich teaching on worship and intercession that could change the destiny of cities and nations. I found the book fresh in insights, experiences and compelling in its urgency. Bob stretched me with his long term vision by advocating that the church should disciple people now for ruling the nations with Christ in the age to come! With this comprehensive teaching the Lord is taking us to another level. Bob has heard a new sound and together with Jill and her 'Sounds of Heaven' CD, we are inspired to hear it too.

Pastor David Carson
Director, Intercessors For Canada
Vancouver, BC

Bob Misst is a humble intercessor, who spends hours on his face, crying out for his own nation, and the nations of the earth. The Lord has commissioned him to write this book, based on his experience of being caught up into the Throne Room of God, where he observed and participated in the heavenly worship of the angelic beings, and millions of the redeemed. In this book, Bob has given us a manual for effective worship, warfare and intercession for these last days.

Intensely researched and full of wisdom and revelation, this powerful and comprehensive book shows us the pattern of Heaven's declarations, multiple choruses, and intercessions. By knowing this pattern, we can join with heaven, as we worship on earth. The book includes practical guidance for worship leaders and songwriters who desire to create the sounds of Heaven.

I heartily commend this book to you. Our Heavenly Bridegroom is calling us into a new depth of adoration and prayer, from which we will never recover.

Jill Shannon
Author, Speaker, Singer/Songwriter
www.coffeetalkswithmessiah.com

Bob Misst has been my teacher in the realm of intercession since 1993. Bob has years of experience in prophetic intercession for the nations. I witnessed the tremendous revival in New Delhi, after his team conducted strategic spiritual warfare over this city.

As intercessors, we are not merely called to pray for our near and dear ones, but to be partners in God's work of salvation in the world. This book gives us the "vision of heaven" for our worship and intercession, whose ultimate goal is to intercede for the glorious second coming of our Lord Jesus, the King of all the earth.

This book is the fruit of years of prayer and prophetic intercession, and I strongly recommend it to all those who are in serious pursuit of their faith. I am confident that Bob's anointed teaching and testimony will raise up true worshippers and fervent intercessors all over the world.

Cyril John
Chairman, International Catholic Charismatic Renewal Services Sub-Committee for Asia-Oceania (ISAO), Singapore &
Vice-President, International Catholic Charismatic Renewal Services Council, the Vatican

"Worship, Warfare and Intercession before the Throne of God" presents a real and powerful perspective on the call and ministry of intercession. During Bob Misst's years of intense, intercessory prayer work, he and his prayer teams have seen the Lord's miracles, in response to their heart's cries, which were aligned and joined with Heaven's purposes. This book is absolutely crucial to our preparation as the Lord's pure and spotless Bride.

As you read Bob's holy testimonies, your own worship, warfare and intercessions will be accelerated in the Spirit, and you will be in step with the very heartbeat of God. You will also experience the Lord's heart of relentless pursuit to dwell with us on earth in response to His Bride's cry for His return. This book, along with the creative partnership of prophetic songwriter and psalmist, Jill Shannon, will fine tune your heart to the myriad of symphonies and choruses of Heaven. Our Bridegroom will rejoice to hear your voice in unison with His Spirit and Bride! What a beautiful way to be adorned and made ready for the Wedding Supper of The Lamb!

Cathy Minnick
Sr. Chaplain
Royersford, PA

TABLE OF CONTENTS

Foreword

We are living in the most remarkable and critical generation in human history. We will be privileged to see the last days' events unfolding, which the saints and prophets, that great cloud of witnesses, have longed to see. I believe that this generation will see the glorious and devastating return of the Lord Jesus, and preceding His return, we will see the Lord restore all that was lost to His church (see Acts 3:21).

Much has been lost since the original Jewish apostolic community died off, and the professing church slid into compromise and man-centeredness for eighteen centuries. One of the most vital patterns that we lost was that of the true worship, which the Father seeks from His covenant people on earth. Moses was instructed to build the earthly tabernacle according to the pattern of the Heavenly Tabernacle, which the Lord showed to him. In the same way, the Lord desires us to pattern the worship on earth according to the worship of Heaven.

Bob Misst is a humble intercessor, a man who spends hours on his face, crying out for his own nation, and for many other nations of the earth. The Lord has chosen him, among other forerunners, as an instrument of restoration to this generation. Their message is as one crying in the wilderness, "prepare the way of the Lord, make a highway of righteousness for the King in your heart," for the time of restoration is at hand. The gracious Lord has taken this prayer warrior into the glorious Throne Room of God, where he has been blessed to observe, absorb and participate in the heavenly worship of the angelic beings and the millions of saints who adore the One who sits on the Throne.

This book combines Bob's throne room experience with his life lessons, learned from many years of intercession and warfare for the nations. Bob has given us a manual for effective worship, warfare and intercession for the last days. To make it even more practical, Bob commissioned me to pattern the worship of heaven on a new worship CD, called *"Sounds of Heaven."* As he was writing this book, I was writing and recording these new songs. It has been an awesome experience, working with him on this duet project. Glory to our God!

For too long, the church has been worshiping "in its own way," choosing lists of songs we like, worship bands that appeal to our soul nature, and structuring our church services around the needs of man, rather than around the burning desires of the One seated on the Throne. Do we truly desire for the Glory of the Lord to come into our sanctuaries, as He did for King Solomon, with the Glory Cloud so weighty, that the priests could not enter the room? If your heart says, "Yes, Lord! I desire the real You," then you will be captivated with this book.

In *"Worship, Warfare and Intercession Before the Throne of God,"* Bob invites us to experience with him the worship of heaven, and gives us prophetic teaching, practical guidelines, and biblical foundations. He shares wisdom and revelation that could only come from Heaven's corridors. In addition, the most wonderful gift that Bob has given us is the pattern of Heaven's anthems, declarations, multiple choruses, and responsive prayers, by which we can join with heaven as we worship on earth.

Beloved child of God, Heaven has drawn breathlessly close to earth, like an unexpected kiss. Your Heavenly Bridegroom is calling you into a new depth of adoration and prayer, from which you will never recover. I heartily and joyfully commend this book to you. May you be lost in heavenly worship and intercessions, until His pierced feet stand, once again, on the Mount of Olives.

Jill Shannon
Author, Speaker, Singer/Songwriter
www.coffeetalkswithmessiah.com

Introduction

The Prophetic Word and its Fulfillment

In November 2009, the National Days of Prayer in New Zealand was held at a beautiful place called Arthur's Pass. Arthur's Pass is the link through the Southern Alps of New Zealand, connecting the East Coast with the West Coast of the South Island. At the heart of this time of prayer for New Zealand and other nations, were the prayer burdens and prophetic words that were released. The key prayer concerns and prophetic mandates were these:

1.) We were called to pray for Israel and for our new Prime Minister, Mr. John Key, who has Jewish ancestry. The need for the nation to be "grafted into the Olive Tree" was highlighted prophetically. At that point in time, the Israeli Embassy in New Zealand was closed. During the National Days of Prayer (NDP), 2009, the Holy Spirit caused me to intercede for a good bit of the night for the two nations, Israel and New Zealand, with deep groans in the spirit and often encountering spiritual battles. The next morning, Warren Lyons, the founder of the NDP in New Zealand, asked me (the author) to lead the morning worship, and I shared that I had been constrained by the Holy Spirit to pray through the night for Israel and our nation, New Zealand. This opened a way for the entire body of intercessors and prophets to intercede for Israel and New Zealand to be related internationally and politically.

2.) We realized by the Spirit, that New Zealand was "passing through" a dark period of change, and so too, was the church (see picture on Arthur's Pass). The key scripture given to us was from Hebrews 12:26-29.
3.) Warren Lyons received a vision of New Zealand as the "Paua Shell." This unique shell is found only in New Zealand (see picture). It reflects many colors (our nation has many ethnic groups), and has a glorious brilliance (reflecting the glory of God).
4.) As prophets and intercessors for the nation, we did not have the wisdom how to pray through this time of international crisis. We needed to seek the Lord every step of the way, through this period of change.

Early 2010, New Zealand re-opened its ties with Israel and re-established the Israeli Embassy. This was a fulfillment of prophetic intercessions, which greatly encouraged us. This prompted the leaders of various streams of prophetic intercessors for New Zealand, to come together and network more closely, due to the urgent sense of the times we are living in.

At that meeting in February 2010, the Spirit of the Lord changed the agenda through a prophetic word, asking the leaders to make a covenant to pray for Israel. Three prophecies, at that meeting confirmed the call to covenant prayer for Israel. We saw the reason for this in the next few weeks. The Gaza "flotilla" problem suddenly raised its ugly head, which brought the world to the brink of another Middle-East war. Fervent intercession went up for this, from the prophets and intercessors in New Zealand (we were covenanted to pray for Israel), as well as from around the world, we saw things change.

A New Zealander activist for the people of Gaza, who was on the flotilla, was hurt, and this caused the NZ media to put Israel in bad light. There was a lot of pressure on the new Israeli Consulate, to shut it down. A New Zealand TV channel continued to highlight the flotilla issue in a way to incite sentiments putting a demand on the fledgling Israel consulate in NZ, regarding the Gaza blockade. I found well-researched articles on the flotilla issue from people in high places in the US universities and US security councils. I sent these articles to the TV channel director, saying that there was not enough research done on their programs, and that the broadcasts were biased. I mentioned that as a news agency, they should cover all the sides of the story and not just one person's opinion. I mentioned that I would write to the broadcasting authority about this. The TV channel stopped broadcasting the flotilla issue quite soon after. I am sure other intercessors prayed for this TV channel to stop the broadcast, as it was not well researched and presented New Zealand as an anti-Semitic nation. This was an answer to much prayer.

The leaders soon realized how quickly things were happening. We had little wisdom how to intercede for all of these international events. We planned a weekend in September 2010, without any agenda, other than to worship our God and to listen to Him. We were to met at Arthur's Pass scheduled for the weekend of September 17th. We had met in November 2009 at Arthur's Pass, "a place where God had spoken to us before". Then, suddenly, on September 4, 2010, the city of Christchurch had a 7.1 earthquake (Arthur's Pass is part of Christchurch domain). Miraculously, not a single life was lost! True, the earthquake caused billions of dollars worth of damage, but there was not one death, nor any serious injuries. Prime Minister Key said, "It was a miracle," when he went to the city and saw the extensive damage to buildings and roads that were split into two. Indeed, only God could have handled this without loss of life.

Despite a snowstorm at the weekend of September 17, 2010, about thirty prophets and intercessors were able to respond to the call. We came together (at Arthur's Pass) to "sit at the feet of Jesus." There was no agenda, other than to seek the heart of our God in worship.

God's response to that weekend of seeking His heart is what this book is all about. At the NDP in November, 2010, held in my own city of Auckland, I presented what I had seen in the visions of the worship in heaven, which the Spirit of the Lord had shown me at Arthur's Pass. I will share this visionary experience in depth, in chapter one. I led the assembly of people into that vision of worship. We had spent the previous evening casting down our idols, The worship that followed was like an explosion. We worshipped from

2 p.m. to 7 p.m., non-stop. The entire weekend bathed in worship, patterned after the worship in heaven.

This book, along with Jill Shannon's CD, *Sounds of Heaven,* is our two-fold offering to the Bride of the Lord. Together, the book and the CD helps us to pattern our worship and intercession as it is in heaven. Jill is a treasure, she is sent by God for this work, bless her. She is a gift to the Body of Christ, by her deep insight into the things of the Spirit, in music and in teaching. We are entering into a season of intense warfare. God is calling us to join heaven as one voice in worship, warfare and intercession, which will release the power of God on the earth in a new way. No eye has seen, nor ear heard, the awesome things to be unleashed by God in heaven for the restoration of His earth. I am sure that there are a number of groups, which are already into this pattern of worship and intercession. All over the earth may the song of the Lamb and the voice of the Bride come forth: "Worthy is the Lamb, the Lion of the tribe of Judah, for He has taken His place and has begun to reign! Hallelujah!"

My experience of worship and intercession in heaven is in complete agreement with the heavenly worship described in the Book of Revelation. Therefore, I share much from Revelation in this book, using the revealed heavenly patterns to inform our spoken prophetic words, and to direct our intercessions. One cannot write a book of this nature without rightly presenting some theological points, which are essential for the Bride to align her heart with the Lord's purposes in these last days.

The theology presented here is "pre-Millennial," which simply means that the Lord Jesus has not yet returned to the earth in His glorified body. He has not yet overturned the kingdoms of this world and begun to reign on the earth.

The Lord warned us that in this world, we must expect tribulation, and that we should not be surprised at the fiery trials that will come upon us. Therefore, let us be prepared for the difficult times ahead. If our hearts are prepared, through intimacy with the Lord now, we will not be offended when these things happen. One thing is certain: The Lord Jesus Christ will return to the earth as the only King, and He will rule over the nations with justice and righteousness. Come quickly, Lord Jesus!

The heart of this book's message is to encourage the church to join with heaven, as we find ourselves in the final days leading up to "the marriage feast of the Lamb." We must pattern our worship, warfare and intercession as it is in heaven, before the throne of God.

Loving blessings,
Robert Misst
New Zealand

Photograph by Robert Misst

The beautiful "Paua" Shell. Some say that the many colors displayed by the Paua Shell, is a reflection of God's Glory in nature in New Zealand.

Photograph by Robert Misst

PART I

The Pattern of Worship

CHAPTER 1

PROPHECY
We See in Part, We Prophesy in Part

The spoken prophetic word, declared under the anointing of the Holy Spirit, is a very powerful tool in discerning how to pray. In this chapter, I will share some of the prophecies spoken at the "Sitting at the Feet of Jesus" retreat in September 2010, held at Arthur's Pass, New Zealand. I will also share some of the prophecies by some other groups that could not make it to Arthur's pass, but were with us in spirit and in prayer.

When a prophet speaks God's Word, he or she is merely the messenger of the Word of the Lord. Some of the people who gave these words have requested to remain anonymous, and I have respected their wish. **In the midst of these many Scriptures and prophetic words, I had an extraordinary visionary encounter, which will become the very core of this book's testimony and biblical teaching.**

The Apostle Peter encourages us to heed the prophetic word of the Lord, and Paul reminds us that our knowledge and prophecies are not yet perfect:

"So we have the prophetic word made more sure, to which you do well to pay attention as to a lamp shining in a dark place, until the day dawns and the morning star arises in your hearts." (2Peter 1:19).

"For we know in part and we prophesy in part; but when the perfect comes, the partial will be done away." (1Cor 13:9-10).

Prophecies and Heavenly Vision from Retreat, September 17th to 19th, 2010

"Behold, I will do something new, now it will spring forth; Will you not be aware of it? I will even make a roadway in the wilderness, rivers in the desert." (Isa. 43:19).

"You cannot enter the new with 'old baggage.' I will strip you of everything that you know that was "good" in the old. You must be changed, from the way you did things before, or the way you thought about things, secular and religious, to enter into the new. Are you willing to be stripped of the old ways? It will be a hard and difficult transition, but you will need to be stripped of the old ways, to enter into the new."

The Vision of Heavenly Worship

On the second morning, we had just started singing our first song of worship, when I (the author) was caught up into a heavenly vision. Suddenly I noticed that the sounds in the room were no longer audible to me. I could not hear the music, or the people singing. I was being escorted from a place of darkness, almost as if it were a tunnel, towards a bright light. I felt the presence of an angel with me, escorting me towards that light. I felt the angel's hand on my right shoulder, which meant that the angel was on my left side. In the dark "tunnel," I could not see the face of the angel.

We moved towards the light, and as we came closer, I began to see what was happening there. I first noticed these huge angels, who were flying around an area that was vast in size, the ends of which I could not discern. The size and countenance of the angels brought the fear of the Lord upon me; there was something so holy about them. I suddenly realized that I was so sinful, and my garments were so dirty, that I could not be in this place. I tried to move backwards, hoping that these huge flying angels did not notice me in this holy place. The angel who escorted me to this place, held my shoulder, preventing me from taking a step back. I sensed his reassurance. When I hear God's heavenly beings speak to me, I hear it not with my ears, but in the center of my head, like a spiritual download straight to my brain! I now knew that I was in the heavenly courts of the temple of God.

While I perceived many things happening simultaneously, I can only write about them, one experience at a time.

First, I will share about the various groups that I saw in heaven. I saw many interesting objects that were able to reflect sound. I will talk a little bit on this. Finally, I will reveal the purpose for which God gave me this vision.

This whole place radiated with exuberance, abounding joy, great expectation, and with the holiness of God. I saw these huge flying angelic creatures, as I saw them I was afraid of their countenance, as they sang, "Holy, Holy, Holy is the Lord, who was and is and is to come." The escorting angel said to me that these are the Cherubim. They are the guardians of the Mercy Seat of God.

I then noticed a small group of people in very stately attire, who were at some distance below the huge, flying angels. I did not count their number, but later, the escorting angel explained this to me. Something about this group made me want to cry. Their whole body language spoke to me of their holiness, love, humility, caution in approach, a passion for whom they serve, and the dedication in their service. They wore crowns on their heads. In the center of this circle of people in beautiful robes, was a bright light, and I could see a throne, almost in outline, as I tried to peer into this beautiful bright light. The escorting angel said, "These are the twenty-four elders."

Beyond this circle of people in stately attire and with crowns on their heads, was a vast group of people, too numerous even to estimate. The way these people were grouped caught my attention. They displayed something so unique, and so special about them. It was like a special identity they carried, a richness about them, and their particular attributes, which were like a gift which they brought with them, to offer to God. However, what awed me is that their physical demeanor spoke of deep love, holiness, and passion for the One whom they worship. The escorting angel informed me that these are the tribes, nations, and tongues, from among the peoples of the earth.

I was awed by the myriads of angels just above them, but lower than the Cherubim. Their numbers looked like millions times millions, and they created a holy anticipation of the One that was to come into their midst.

Seeing all this kept reminding me of my dirty garments, as compared to the stately robes that this holy assembly was adorned with. I wanted to back off from this place, with the hope that no one has noticed me. The escorting angel pressed my shoulder, to let me know he was with me, and not to be afraid. He was very assuring, and I knew that I needed to be there, and that his assignment was to escort me and show me around.

I was caught up by strange shapes of the most exquisite beauty, the likes of which I have not seen on the earth, not even in pictures, or in artistic impressions. The escorting angel told me that the heavenly temple filled with objects, have the ability to resonate with heavenly music, and create auroras of color based on the harmonic sounds that come from your voice. I hummed a short tune to try this out, and I was amazed at the sound that matched the tune of my humming and played a number of beautiful, intersecting improvisations.

Suddenly, I was able to hear the worship in the holy heavenly temple. If I had been in my mortal body, I could have died, just from the beauty of the sound.

The Cherubim sang their song: "Holy, Holy, Holy is the Lord." I began to notice that each of the groups in the heavenly temple had their own song or chorus. I was hearing multiple choruses all at the same time, yet in perfect harmony. Added to these multiple choruses were the perfect musical harmonies played by these beautiful instruments, which I have never seen on the earth. I was so awed beyond words, basking in the most glorious worship in heaven. There are other earthly sounds that hit my ear, depending on the choruses, the peels of distant thunder, flashes of lightening, and the sounds of wind and water.

I then saw from the throne a Person coming, with the most brilliant attire. I was told by the escorting angel that He is the Lamb of God. I was filled with the fear of the Lord, and almost did not want to look, for in His entire being, the exuberance of love and mercy just shone forth, like no star in heaven ever could. The twenty-four elders literally tore away the crowns from their heads, and cast them on the ground. They prostrated themselves before the Lamb of God. He moved with such gentleness, love and mercy towards the elders, and lifted them to their feet.

As each elder gazed at the Lamb, each of their gazes held the most profound expressions of love, honor, respect and holiness for one another. The Lamb gently picked up the crown, to place it on the head of the elder. I was permitted to hear the elder say, "No Lord, only You are worthy to wear the crown." I could see in his expression, the elder's love for the Lamb, and that it was far more important than the crown, and all it signified. After the Lamb put the crown on all the elders' heads, He asked one of the elders, pointing to the vast group of people from the tribes and tongues and nations, *"Who are these people?"*

The elders replied, "These are those who washed their robes in the blood of the Lamb, for they loved not their lives." The Lamb of God moved to one of them, and placed a crown on his head. The person refused to accept the crown, saying, "Only You are worthy Lord, to wear the crown." This person then said to the Lamb of God, "There are many on the earth that still need to be saved, so we are not worthy to wear the crown." The Lamb then assured the person, *"As you worship Me, I will set the people on the earth free."* The Lamb embraced that precious person, and His embrace touched me so deeply, in the way that He expressed His holy love, honor and respect for him.

The escorting angel began to speak to me about the worship in heaven. He taught me that worship in heaven has a distinct purpose. Each chorus sung by each group has a purpose. The worship in the heavenly temple advances the end-time purposes

of God on earth. Each chorus sung by each of the seven groups I was shown in the heavenly temple, has a message for the church.

I then saw the Lamb taking the scroll and breaking the first seal. I had no idea what was happening. I did not connect it, at that time, to Revelation, chapters 4 and 5. The worship was still in a deep place, and the twenty-four elders were now holding golden bowls of incense, as they were praying. I could not distinctly hear their prayers. The escorting angel explained that they were urging the Lamb to take the scroll and open its seven seals. The Lamb broke open the first seal, and there was a great sigh, and sounds that sounded like holy cheers from the tribes, tongues, and nations.

Almost immediately, one of the mighty Cherubim called out in a thunderous loud voice, which shook the very atoms of my body. He said, "Come forth," and a white horse with an angelic rider presented itself before the entire assembly. The Lamb spoke to the rider, and sent the angel forth with a great command to proceed with his holy assignment. I was allowed to see four seals broken by the Lamb amidst worship, intercession and warfare in the heavenly temple. Four angelic beings rode forth on the horses. The first was white, the second was red, the third was black and the fourth was on a pale green horse. I was not permitted to hear their assignments, but was assured by the angel that I would be told shortly.

Suddenly, the lights in the heavenly temple seemed to go out, and I was brought back to the prayer room. The vision seemed like all eternity had passed, and I saw the faces of my brothers and sisters staring at me, asking me what vision I had seen.

I received the same vision the second time, the next day. This occurred as the team began to worship the Lord in intercessory worship, bringing our "bowls of incense," the intercession for the people of our nation, and the nations of the world. I saw the Cherubim sending out angels in thousands, and they were returning with harvested sheaves in their hands (representing the reaping of the harvest of the people, whom we carried with us in worship).

This vision was confirmed by a reading, almost immediately, from Psalm 126.

"*When the Lord brought back the captive ones of Zion, we were like those who dream. Then our mouth was filled with laughter and our tongue with joyful shouting. Then they said among the nations, 'The Lord has done great things for them.' The Lord has done great things for us; we are glad. Restore our captivity, O Lord, as the streams in the South. Those who sow in tears shall reap with joyful shouting.* ***He who goes to and fro weeping, carrying his bag of seed, shall indeed come again with a shout of joy, bringing his sheaves with him***. (Psalm 126:1-11)."

As I continued to have this heavenly vision, I heard the Lord asked me a question: *"To what will I compare this group*?" (see Luke 7:31)

I said, "Lord, I don't know the answer."

I was then given the vision of the shepherds guarding their flock, and the Angel appeared to them with the message of the birth of the Lord. While this passage relates to the first coming of our Lord as a child, this passage now spoke to us about His second coming, which is to take place soon.

Then the vision ended.

Luke 2:8-20 gives an account of this event in the Scriptures. Important elements that jumped out of the passage were these:

- It is about the coming of the Lord (however, we are to relate it to His second coming).
- Luke 2:8 speaks about "keeping the night watch." This could mean "intercession." It is a time to intercede for the nations and the church. During this time, there will be important visitation of angels who will bring us "good tidings of joy." In verse 13, the "suddenly" event takes place: the angels that go out with good tidings will minister to the heirs of salvation (Heb.1:14).
- The passage speaks about the city: God cares for and loves the city.
- It calls the Lord's disciples to make haste to go to the cities with the message (see vs. 16).

- The message is about the events that must take place heralding His coming: shakings, stripping, and salvation. Make way for the return of the Lord.
- The people in the cities will be amused at our message, as they were when Noah was building the ark.
- However, we are to return to the feet of Jesus, glorifying and praising God.
- Humility: the shepherds speak of people of humble stature. Humility is the key to doing His end-time works.
- There was a prophetic word about the great eagles (representing the leaders first) that soar in the high places in the presence of God; they must come down to the lower rocky clefts of the mountains, to be stripped of their old feathers. The vision spoke of eagles with outstretched wings, their feathers stripped off their bodies by mountain winds, and a sense of their nakedness. They must lose their old feathers, talons and beak, and must wait for the *new glorious feathers to grow*, so that the eagles can soar even higher into the presence of God. We must come into an understanding of the end-time purposes and ways of the Lord. This will require a humble spirit by leaders.

The founder of the National Days of Prayer, Warren Lyons, gave this prophetic word:

"I will walk with you in the snow as I walked with Adam in the garden. My Righteousness is like the great mountains. The Snow is likened unto My garment of righteousness; I will walk with you in the snow as I walked with Adam in the garden. My righteousness will prevail upon you and upon the Nation. I AM covering you like a blanket of snow, your iniquities and sin I will remember no more. This is a day of atonement, even as you have drawn close to Me, My righteousness I have given freely to you.

"I see the desire in your hearts to worship Me; I will open the doors of heaven to My people, to enter into worship before My throne. A new understanding will be revealed of who I AM, and the glory of

My presence for you is to be seated in heavenly places with Me. I have called you Holy Nation and a Royal Priesthood, to minister before My throne. I have called, chosen, cleansed and sanctified you to worship before Me, casting your crowns in service to your King, as this is also the Father's will, and the will of the elders before My throne.

"Even as I have covered you with a new garment of righteousness before Me, I am at work in your hearts to release you into a new realm, the heavenly realm and the desire to worship before Me. I am causing your hearts to be broken and to be stripped of the old that has hindered you, into free worship, and to prepare your hearts, so the new may come forth as Holiness unto the Lord. A Royal Priesthood accepted by the Father. Allow My purifying fire to come afresh into your hearts, that will set you free to worship in the Heavenly Realm, and to know, see and experience the throne of worship. Amen. He alone is worthy. Jesus is our KING. Praise the name of Jesus all over the earth, and let us bow down and worship Him."

Warren Lyons explains Bob's vision of heavenly worship, warfare and intercession

"As you Worship Me (the Lord), I will give you a new understanding of who I AM, and the glory of My presence. You are a royal priesthood, a holy nation. When you come before Me in worship, cast your crowns before Me." God is going to prepare us, to cut away those things we have held on to that He wants us to give up, so that we might worship Him in spirit and truth.

The key word in understanding worship is humility. We must be willing to be stripped like the eagle. The willingness is our free will, our desire, a desire that the Lord will do the stripping of our hearts, minds, and concepts; this is true humility. Let the purifying fire of the Holy Spirit come into your heart to burn all that is not of Him, to bring you to true holiness. Jesus' righteousness will reign on the mountains (Isaiah 2:2-3). Jesus will cover your sins as a blanket of snow; Jesus will walk with you in the snow, as He walked with Adam in the garden.

Other prophetic words and Scriptures received at the Arthur's Pass weekend

"As I shake this nation (NZ), I want you to trust Me with the shaking. Intercession will put forth a canopy of mercy over the nation. Trust Me in the shaking of all nations. I want you to agree with Me in My shakings. I will break their stony hearts and give them a heart of flesh."

"I see a picture of Moses, raising his hands supported by Aaron and Hur as Joshua battled with the Amalekites. Right now, a battle rages in the land for the hearts of men, and the call goes forth for repentance, the turning to God and to the fear of the Lord. There are coming even more shakings of the land, in the government, and even in the churches: everything will be shaken, so that only that which cannot be shaken of the Lord will remain. Lord wash us, and make us as white as snow."

"Prepare the way for the coming of the Lord. Prepare, for the Lord is preparing a place for us."

A person had a vision of a page-boy, a town crier, as it were, that went around in the city of Christchurch, shouting, "Prepare the way of the Lord!"

Song of Songs 8:6 was read: *"Put me like a seal over your heart, like a seal on your arm. For love is as strong as death."*

We heard the reading about John the Baptist: "Repent, make straight your paths, and bring forth fruit worthy of repentance."

We read from Revelation 5:8: *"And when He had taken the book, the four living creatures and the four and twenty elders fell down before the Lamb, having each one a harp, and golden bowls full of incense, which are the prayers of the saints."*

This scripture speaks about intercession in worship. This book is an encouragement to those who are part of the "Harp and Bowl" ministries, and it is my prayer that it will take them to an even deeper level.

We read from Hebrews 12:1-2: *"Therefore, since we have so great a cloud of witnesses surrounding us, let us also lay aside every encumbrance and the sin which so easily entangles us, and let us run with endurance the race that is set before us, fixing our eyes on Jesus, the author and perfecter of faith, who for the joy set before Him endured the cross, despising the shame, and has sat down at the right hand of the throne of God. For consider Him who has endured such hostility by sinners against Himself, so that you will not grow weary and lose heart."*

We heard from John 15:16: *"You did not choose Me but I chose you, and appointed you that you would go and bear fruit, and that your fruit would remain, so that whatever you ask of the Father in My name He may give to you."*

"Love Me, seek Me, it's a time to be real and honest. I will make this land My own. Will you bend the knee to Me, trust Me in this time?"

"Now is the time for the new thing that the Lord is doing. A word has been given: *'I will make all things new, yet you revert to the old. You need to be transformed by the renewal of your mind (see Rom. 12:1). Come to Me empty.'*"

Vision: A pen signing off documents of the nations (false documents). Instantaneously, there was an earthquake. Then I heard a voice say, *"I AM the God of love, justice and righteousness and holiness."*

"Blow the Shofar, sound the alarm, let it shake the nations and bring the Fear of God that the nations have cast aside and do as their please. My shakings ought to bend the knee and bring contrite hearts before Me."

Message in tongues with interpretation: "Prepare the way, Come Lord Jesus. I am coming soon."

"Go out as Lambs among the wolves. Be near the feet of Jesus, as lambs that have been stripped."

"May the eyes of His church be only for Him (see 2 Chron. 16:9). For the eyes of the Lord move to and fro throughout the earth that He may strongly support those whose heart is completely His."

We read Revelation 5:1-14:

"I saw in the right hand of Him who sat on the throne a book written inside and on the back, sealed up with seven seals. And I saw a strong angel proclaiming with a loud voice, 'Who is worthy to open the book and to break its seals?' And no one in heaven or on the earth or under the earth was able to open the book or to look into it. Then I began to weep greatly because no one was found worthy to open the book or to look into it; and one of the elders said to me, 'Stop weeping; behold, the Lion that is from the tribe of Judah, the Root of David, has overcome so as to open the book and its seven seals.'

"And I saw between the throne (with the four living creatures) and the elders a Lamb standing, as if slain, having seven horns and seven eyes, which are the seven Spirits of God, sent out into all the earth. And He came and took the book out of the right hand of Him who sat on the throne. When He had taken the book, the four living creatures and the twenty-four elders fell down before the Lamb, each one holding a harp and golden bowls full of incense, which are the prayers of the saints.

"And they sang a new song, saying, 'Worthy are You to take the book and to break its seals; for You were slain, and purchased for God with Your blood men from every tribe and tongue and people and nation. You have made them to be a kingdom and priests to our God; and they will reign upon the earth.'

"Then I looked, and I heard the voice of many angels around the throne and the living creatures and the elders; and the number of them was myriads of myriads, and thousands of thousands, saying

with a loud voice, 'Worthy is the Lamb that was slain to receive power and riches and wisdom and might and honor and glory and blessing.'

"And every created thing which is in heaven and on the earth and under the earth and on the sea, and all things in them, I heard saying, 'To Him who sits on the throne, and to the Lamb, be blessing and honor and glory and dominion forever and ever.' And the four living creatures kept saying, 'Amen.' And the elders fell down and worshiped."

Words Received the Same Day from Other Prayer Teams in New Zealand that Supported the Group at Arthur's Pass

The following quotes are all taken from the testimonies of other intercessors around New Zealand, who also "Sat at the Feet on Jesus," on the same day, but in other locations. Due to the snowstorm, these groups could not make it to Arthur's Pass.

"We read from 2 Samuel 15:1-19: The story of Absalom, the son of King David.

"In short, Absalom was the perfect specimen of manhood. He was the natural son and heir, and he had stood beside the way to the city gate over a period of time. With flattery and promises, he beguiled and persuaded many people entering Jerusalem with a legal dispute, so that they would follow him. Verse 6 tells us: "*So Absalom stole the hearts of the men of Israel.*"

"He organized a coup and entered Jerusalem. His father, King David, and his followers fled the city for their lives. Eventually, Absalom was killed (see 2Sam 19:9) which left all the tribes in strife, saying, "*The King saved us out of the hand of our enemies, and he delivered us out of the hand of the Philistines. In addition, Absalom, whom we anointed over us, is dead in battle. Now therefore, why do you say nothing about bringing the King back?*" (See vs.10).

"The parallel is that we, the offspring, have taken over our Father's house one way or another, according to our church traditions and/or programs, and He wants it back!

"The following dream was received by a prophetic intercessor. In the dream she was holding a baby boy, aged 18 months. The child was wanting her attention by asking for cuddles. As she was cuddling him, her other son, aged ten years, approached her, also asking her to cuddle him. So she did, as well as giving him a kiss on the cheeks. Not to be left out, the other two sons aged 15 and 20 years, wanted to be cuddled and kissed. In the dream there was utmost respect between the mother and her sons. Not only that, but between the boys, there was no sense of rivalry, animosity or jealousy. They all knew where they stood in the family ranking, and embraced, honored and encouraged each other. [End of dream.]

"The interpretation of the dream was this: The Heart is in the Home. This represents a new way of being "His family." This speaks of the presence of God coming so powerfully among His people, that we will all know that we have value. We will know that we are graced with His love and favor to such a degree, that we will not be jealous of one another; rather, we will nurture, support and encourage one another.

"They prayed for the lack of unity in the house of the Lord, the church. The people stood in agreement, crying for the children to be saved and brought home. There was a childbirth-like travail. **In view of all the above, we felt we needed to invite the King back to His house.**

"They reported this: "We have been challenged by Luke 1:19, where Gabriel describes himself as one who stands in the Presence of God. His major assignment was simply to stand in God's Presence. Then he was sent to speak the Word of God and to show His power.

"We read Psalm 82:1-8: *"God takes His stand in His own congregation; He judges in the midst of the rulers. How long will you judge unjustly And show partiality to the wicked? Selah. Vindicate the weak and fatherless; Do justice to the afflicted and destitute. Rescue the weak and needy; Deliver them out of the hand of the wicked. They do not know nor do they understand; They walk about in darkness; All the foundations of the earth are shaken. I said, 'You are gods, And all of you are sons of the Most High.' Nevertheless you will die like men and fall like any one of the princes. Arise, O God, judge the earth! For it is You who possesses all the nations."*

"Verse 5 of this Psalm speaks of the government, the magistrates and judges, the political and judiciary systems. These systems do not know or understand the ways of the Lord; they walk in the darkness of their own complacent satisfaction. All the foundations of the earth, the fundamental principles and infrastructures upon which rests the administration of justice throughout the world, are being shaken and tested for righteousness.

"We sang a song whose words seemed so prophetic to us: 'O Lord, You are beautiful. Your face is all I seek. For when Your eyes are on this child, Your grace abounds to me.'"

"We spoke of redeeming the time, from Ephesians 5:15-21. The focus was especially on verse 19: 'Speaking one to another in psalms and hymns and spiritual songs,' and on verse 20: 'Giving thanks always for all things in the name of our Lord Jesus Christ to God, even the Father.'"

"Colossians 4:5 was highlighted: Walk, conduct yourselves with wisdom toward outsiders, making the most of the opportunity.

"It matters how we spend each hour, each minute. We had a sense of how critical is the hour we are in, and how short the time we have left. How we spend this time will make all the difference in how we come through the trials that will come to pass shortly.

"We read Psalm 84:1-12:
'*How lovely are Your dwelling places, O Lord of hosts! My soul longed and even yearned for the courts of the Lord; my heart and my flesh sing for joy to the living God. The bird also has found a house, and the swallow a nest for herself, where she may lay her young, even Your altars, O Lord of hosts, My King and my God. How blessed are those who dwell in Your house! They are ever praising You. Selah.*
'How blessed is the man whose strength is in You, in whose heart are the highways to Zion! Passing through the valley of Baca they make it a spring; the early rain also covers it with blessings. They go from strength to strength, every one of them appears before God in Zion. O Lord God of hosts, hear my prayer; give ear, O God of Jacob! Selah.
'Behold our shield, O God, and look upon the face of Your anointed. For a day in Your courts is better than a thousand outside. I would rather stand at the threshold of the house of my God than dwell in the tents of wickedness. For the Lord God is a sun and shield; the Lord gives grace and glory. No good thing does He withhold from those who walk uprightly. O Lord of hosts, How blessed is the man who trusts in You!'

"This psalm reminded us of how our hearts yearn to be in the courts of the Lord, and what a privilege it is to praise God in His courts. Our hearts are on a pilgrimage. God bestows good things on those whose walk is blameless. Blessed are those who trust in Him.

"Isaiah 62 spoke to us of how for Zion's sake, the Lord would not remain silent until her righteousness and salvation were like a blazing torch, until her righteousness would be evident to the nations, and that she would be called by a new name. He would rejoice over her, as a bridegroom rejoices over his bride.

"At this point, one of the intercessor's shared a dream from the night before. She was to marry her husband again, it seemed. She was dressed normally, but put on her wedding garment over her clothes. On the way to the church, she saw two young men dressed in tartan kilts. It seemed that they were from a different "clan/tribe," and that they had been guests at a wedding celebration that had taken place sometime before. Upon entering the church, the bride knew that her bridegroom was standing at her right hand, and she saw her two sisters there. The absence of a brother was noticeable (she has two brothers). The door opened and her elder brother entered, and she remarked upon how much he resembled her father. Suddenly it was her father standing in his place with outstretched arms, calling her to come to him. She ran to him and leaped into his arms, upon which he bid her to "*come up higher,* so that you may see over my shoulder."

"The group agreed in spirit to the desire to see large crowds of people gathering to worship the Lord, of the older women mentoring the younger, bringing healing to the hurt and drawing back those who had strayed away from the Lord. The Holy Spirit then spoke as if in a loud voice the word, "Salvation!" (This was punctuated by another aftershock of the earthquake in Christchurch, New Zealand.) We then read the scripture of the raising of Lazarus, as we pictured the scene of the Lord Jesus commanding the stone to be rolled away from Lazarus' tomb. Martha expressed concern about the state of his decay. He answered her, *"Did I not tell you that if you believed, you would see the glory of God?"*

"A vision was received: God's people helping to remove the grave clothes from those who heeded His call to come out (of the "old" lifestyle). We were to bring God's glory into the darkness that covers the earth. Some would be drawn to His glory, others would flee from it.

"God's glory was highlighted in Psalm 24:7. 'Who is the King of glory? The Lord strong and mighty, the Lord mighty in battle.'

"One person received a vision: A blanket of mist came down. This was God's glory, which gently lay, and was spreading over the Canterbury Plains. It settled, bringing peace. Next, it permeated down into the earth, through the cracks, taking His peace deep down.

"Another vision came: of little sand volcanoes (liquid factions) bubbling up from the ground. And a previous vision was shared, consisting of small islands in a swamp, with a flame burning from each. This spoke to us of the unity in Christ, and of the spiritual leaders of the local body of Christ represented in this community, along with the body and the praying intercessors. Praise and prayer was given for the closer connectedness of these four and their God-given purpose.

"We also had a vision of the large, new community hall (being built now), which is now dark, being filled with the light of God's glory. Perhaps this will be used for interdenominational gatherings, overflowing with young people. This vision was received in the past once before.
Prayer was made for the ability to "give up" some of the things of the past, as illustrated in the loss of historical buildings, which were of their own season, and the ability to be freed for the wisdom and the new things that the Lord would have for our communities. We prayed for the deliberations and decisions made by mayors and councilors, to be turned to God's purposes.
One outcome from this praying group is the desire to have a "thanksgiving" service/gathering for the city of Lincoln and surrounding districts of Canterbury, NZ.[1] At the end of our meeting, we asked the Lord to pour out His peace and love upon the land, to heal it, to turn hearts back to Himself, and to still and soothe the unsettled ground under our feet. Our desire is to see the opportunities the Lord places before us, to partner with Him in this work, and to carry His glory wherever we go. All Praise To Him!"

[1] It is interesting to note that American President **Abraham Lincoln** decreed that "Thanksgiving" should become a national American holiday. This celebration was originated by the Puritans, who considered Thanksgiving to be like the Israelite's harvest festival, the Feast of Tabernacles. Perhaps it was prophetic that the intercessors around NZ city of "Lincoln" felt led to celebrate a Thanksgiving service during the season of the Fall Feasts.

The following vision came from a sister, who supported those of us at Arthur's Pass, as she spent the day in prayer and fasting from her home:

"In the morning while praying, I had a vision of a bridegroom at the altar of a great big church. The church was beautifully decorated, all in white, and with pearls all around the inside of the building. In the pews were seated thousands and thousands of Angels. Heavenly music was coming out of their mouths. It was absolutely magnificent. Everybody was waiting with anticipation for the bride to come through the door.

"My vision went into another room, where there was a bride trying to get ready. She was very, very frustrated. She was looking for pieces of her garments, such as gloves, shoes, and her veil. She could not find them. In the room was a huge clock. The clock was ticking with loud clicks. Time was running out. The bride found the gloves and one shoe, but she could not find the veil. She was in confusion and starting to lose hope. She had never prepared herself properly for that day.

"Back in the church, I saw the bridegroom starting to weep. I never saw His face. All the angels started to weep too. Their tears were everywhere. The tears ended up being a raging flood of water, and the next minute, everything was washed away, including the bridegroom, the pews and the Angels. The whole place was empty. Then I noticed in the doorway, that the bride was standing there, crying her eyes out, and still had not been ready. She fell to the ground and cried bitterly." [End vision].

This sister provided these supportive Scriptures for this vision: Matthew 25:1-13; Revelation 22: 17, as well as the interpretation of this vision:

"I believe that the Lord said to me that we are the bride. We are a people of confusion, a people still in apathy, looking for spiritual experiences and answers to satisfy us. We are looking to man, and not to God. We have for too long been satisfied with church, church performance, church entertainment. Churches have been built to satisfy the need of a hungry soul. A hungry soul will be fed, but it does not last. We are satisfied with feeding ourselves on bread which does not satisfy us, but leaves us hungry, and we are drinking water which still leaves us thirsty.

"When Jesus said, *'Eat My flesh and drink My blood,'* He knew that this is what would satisfy us. It is He and only He who will satisfy, not entertainment programs. We need our hunger to be satisfied, we need God's Glory to move in the Church, and we need His manifest presence to come and to stay. When God's Glory comes, it will be shown to the world for them to see. There have been revivals that have come and gone. Good things have come out of them, but bad has also come, such as man building man's own churches.

"God wants His Glory to come, which will last, which will change our hearts and satisfy our spirits. His presence will fulfill the longing which our spirit is looking for. We will not have to run after meetings to entertain us, which leave us hungry and dry. We want something bigger, greater, and more powerful that will remain. We need to seek for His Glory and His Glory only. This is the Glory that no man will take for himself, and which no man can touch, for the man who touches God's Glory will die.

"Wake up, Church, and begin to seek the Lord for repentance, repentance that will change our lives. We must learn to die daily, and to walk through the door of repentance. When death to self takes place, then His perfect will can come into our lives. We are a bride, and we must make ourselves ready for Him. He wants to work in our inner man. Human Flesh and God's Glory cannot mingle. The Lord wants us to set aside our agendas and our own plans, to make room for His plans. Then, when true repentance comes, it makes a platform for his Glory to stand.

"Throughout many periods in the Old Testament, there was much famine in the land. We are a people of famine, but we want to be a people who desire to be fed by the Lord only. We have been fed by man, and man has been attempting to satisfy our spiritual famine, but we are still hungry and we are still thirsty. We need Him and only Him. Only when we have the Lord Himself, will we feel satisfied. There is just no substitute for the One who created us, and knows our needs better than we know ourselves.

"The change in our lives is about our everyday life; it is really about whose will you are following, yours or the Lord's. This is my challenge to the Body of Christ, including myself. We want to see the Lord's Glory, and to be fed by Him. The warning in this pursuit is that as you seek for His Glory alone, it could mean that the Lord could turn your life upside down. However, it will be worth every moment of the discomfort in your lives, for the benefit of seeing the Lord's Glory revealed in His Church, and to see His Bride made ready for her Bridegroom. Then we will have all the necessary bridal garments laid out on the bed for that day."

I, the author, have shared all of these Scriptures and prophetic words with you, because they support the visionary encounter that the Lord granted to me. I hope they have been meaningful to you, dear reader, as they launch us into the deeper worship of Heaven.

CHAPTER 2

VISION OF HEAVENLY WORSHIP – I

Step into the Worship Pattern of Heaven

Background

The Lord drew me into prophetic intercession around the year 1987. Prior to that, I was part of a group of elders that oversaw the pastoral care of a covenant Christian Community in Mumbai, India. It was customary for me to spend a few days towards the end of the year in quiet retreat, mainly seeking the Lord's guidance for the community's New Year plans. At these personal retreats of prayer and seeking the Lord, a "retreat director" guided me, or at times, he would just be around to help me with my own thoughts, scripture readings and discernment.

However, on a personal retreat in 1986, my retreat director was the Lord Jesus Christ Himself. As I entered the room allotted to me for the weekend, I felt the presence of the Lord in the room. I was on my knees in worship. It was around 4 p.m. in the afternoon. When I got up from the floor to earthly consciousness, it was around 11 p.m. in the night. Later, I learned from the retreat director that he had come and gone several times, in an attempt to meet me and to see if I needed his guidance for the time I was to spend at this retreat house. He could not enter the room, even though the door was wide open. An aroma of incense filled the room. The room was filled with the most awesome presence of the Lord.

What happened during those hours was something similar to the Isaiah chapter 6 account. The Lord commissioned me into the ministry of prophetic intercession for the nations. I did not understand anything about this ministry at that time, and the whole experience in the spirit of mighty angelic beings in the room, with incense and golden tongs, was like a dream. Nevertheless, the Lord confirmed that it was not a dream, for when I got up from the floor at 11 p.m. in the night, I noticed that my weekender bag was on the ground, not on the table. It was open, and my Bible was on the floor in the open position, but inverted. Since I was worshipping, I could not recall at any time having opened my bag. My open Bible lay on the floor, I picked it up, and it was Isaiah chapter 6! I said to myself: "This experience of the angels and the incense and the tongs and the coal touching my mouth, this is not a dream." The Lord was calling me into a new ministry of prophetic intercession. At that point, I knew nothing about prophetic intercession; leave alone a ministry to the nations. Being just a pastoral leader, with spiritual gifts that worked through me, like prophecy and occasional healing, apart from being the worship leader for the community when we gathered on Sundays to worship the Lord as a community.

A journey began into an unknown ministry and an unknown destiny, learning to walk by faith, and just knowing that the Lord was in charge of it. I spent time with the retreat director over the next eight months, praying and discerning this call. He advised me that unless I launched out by faith and trusted in God, I would never know if this call was from God. I did just that in faith. From 1987 onwards, it has been the most incredible journey of intercession for nations. The Lord has taken me to many nations; I don't think I could even have dreamed of going there. He has drawn me into learning experiences with some of the most awesome men and women in intercession, whose books I had read and admired. He has shown His power at work through intercessory prayers that has surprised me. I was part of prayer mission teams that went around the world, interceding for the nations. For example, the prayer mission in West Germany was for the purpose of the unification of East and West Germany. This was my first awesome experience of prophetic intercession at work in the nations, when after the prayer mission was done, I read and saw the news that the Berlin wall tumbled down, and the two nations united as one. Before my eyes, prophetic intercession was preparing the destiny of nations for the return of the Lord Jesus to the earth.

Then after the crumbling of the Berlin wall, the prayer experience in Mexico followed. Through this prayer mission, Mexico was freed from years of communist government rule.

Another powerful experience of prayer was when I was with a prayer team directed by Derek Prince in South Africa. A politically instigated avowed blood bath averted through the ministry of the prayer mission. At the government office, while President Nelson Mandela discussed the issues with the Zulu Chief, who had threatened to have a blood bath in the city, if his demands were not met. Intercessors in an adjacent room bowed their heads and bent their knees in prayer for the nation. Again, through prophetic intercession, a nation moves towards its end-time destiny.

In India, I was simply awed to see the ministry of prophetic intercession spread like wild fire. It was a privilege and a humbling experience to intercede with anointed men and women all across the nation. During those ten years of ministry in prophetic intercession in India, it was amazing to see anti-Christian governments change their stance towards the gospel of Jesus Christ. Local governments known to have an anti-Christian attitude invited Pastor Benny Hinn to come and preach the gospel in India. The healing services of Pastor Benny Hinn drew millions of attendees, with numerous healings and dedications to follow Jesus as Savior and Lord.

Over the years, I have seen the Lord do some awesome miracles. The raising of the dead and the multiplication of food, among other miracles of healing, were real faith builders. A young man declared clinically dead, came back to life again, six hours after I had prayed over him in the hospital. This teenager consumed an entire bottle of sleeping pills. He wanted to commit suicide. His family members took a while to realize what had happened and found out why he slept all day. Unfortunately, vital hours passed before admitting him to hospital. At 11 p.m., the day after the incident, my brother-in-law asked if I would come to the hospital and pray for this young man. I obliged.

On entering the hospital room where the young man was, the father, a medical doctor himself, said to me that it was over 36 hours his son was hospitalized, and he had not come out of the coma. The various medical instruments connected to his body showed no signs of his vital organs functioning. He felt that it was no use praying for him, as he was clinically dead already, and should he miraculously come back to life, his brain would not be functional again, and his son would be a "vegetable." He left the room, and I stood by his bed, along with the person who brought me to the hospital to pray for this teenager. I began to pray in tongues, in the realm of the spirit, I was with the Lord, and was asking Him how to pray. After several moments of quietness, I witnessed the resurrection of Lazarus. In response to this experience in the spirit, by faith, I called out the young man by name, saying, "In the name of Jesus, come out."

I then left his room and went home, and fell asleep at around 1 a.m. I was awakened at 7 a.m. by a telephone call. The excited voice at the other end was my brother-in-law, the person who had taken me to the hospital the previous night, saying that the teenager awoke at 6 a.m., and was confused as to why he was in the hospital. He told the nurse to remove all the pins and needles in his body that hooked him up to the various machines, which were indicating the condition of his vital organs. He then walked home. He moved on to finish his graduation from university, praise God no sign of a brain turned into a "vegetable" condition, Praise Jesus!

Another miracle I will not forget was the multiplication of food, while ministering to the poor in Mexico. I had completed a training course in discipleship in New Mexico, USA. The pastor had a ministry to the poor, those who lived across the border in the city of Juarez, Mexico. I went along to minister with him. During the day, I worked alongside the people who were digging and making a road. At the end of day, I had to distribute from a 100-pound bag of beans meant to feed them with food for work. I was asked to give them a saucepan full of beans, which was about four pounds in weight. When the people lined up, I realized that the food in the bag would not provide all 300 people with the generous dollop of four pounds per person. Amazingly, it did, and four pounds left over for the pastor (and me) at the end of it all. The pastor then turned to me and said with a smile, "This happen every week," as if it were a normal thing to expect! My faith journey continues to this day. All glory to our awesome, loving and caring God!

The visionary prophetic experience of being taken to heaven to experience the worship, warfare and intercession in heaven, which I discussed earlier, caused me to "take another look" at why this experience was given to me at Arthur's Pass.

The vision of worship, as it is in heaven, resembled the book of Revelation, chapters 4 and 5. In other words, the Lord was speaking to us (and through us), that our worship had to be realigned, or patterned after, the worship I had experienced in heaven.

"After these things I looked, and behold, a door standing open in heaven, and the first voice which I had heard, like the sound of a trumpet speaking with me, said, 'Come up here, and I will show you what must take place after these things'" (Rev. 4:1).

Duplication of heavenly worship is impossible on the earth, as the resources in heaven are unimaginable, however, we must pattern it as the scriptures bear witness and the Spirit of God beckons the church to follow.

The Midrash – the Jewish Hermeneutical Method of Biblical Study

At this point, I want to bring in an understanding of prophecy, which perhaps some in the ministry of prophetic intercession have heard little about. There is a Jewish hermeneutical method of biblical study called the *'Midrash.'* The Midrash actually interprets prophecy as a cyclical pattern of multiple historical fulfillments, pointing to an ultimate fulfillment of that prophetic word. This Midrashic Jewish concept views prophecy as a pattern recapitulated repeatedly, or thematic reinstatement several times, until it has an ultimate fulfillment.[2]

In the days of Jesus and the new church that was born after Pentecost, it would seem that they employed a Midrashic hermeneutic, as that would have been the most common way first- century Jews, like the apostles, would have been brought up in their understanding of the Old Testament. There is a Greek (Hellenistic) way of thinking, and there is a Hebrew (Hebraic) way of thinking. When Paul spoke to the Jews of the Diaspora (those Jews who were not living in Israel, but who had been scattered to other nations), he reasoned with them, using the Hebrew way of thinking. However, in Athens, when Paul was preaching the gospel at the Areopagus (See Acts 17:22-31), he reasoned using the Greek way of thinking. "Jews seek for a sign, Greeks seek for wisdom. (See 1Cor.1:22).

Pattern not prediction

Greek thought heavily influenced the Mediterranean world and the Middle East for almost three centuries before the birth of Jesus. After the death and resurrection of the Lord Jesus Christ, and after the Jewish apostolic community had died off, the Christian faith was widely disseminated throughout the Greek-speaking gentile known world. The Christian understanding of the Bible became more and more influenced by Greek philosophers and their worldview, until it was completely cut off from the Hebraic understanding that had been handed down from the Jewish apostles[3]. As a result, biblical exegesis even now, looks at biblical prophecy from the Greek point of view, and sees it as a prediction with a onetime fulfillment. However, the Hebraic approach was to view the biblical prophetic word as a **pattern** that existed in the Old Testament, and which had

---

[2] "Midrash" article by James Jacob Prasch. www.moriel.org/articles/sermons/midrash.htm

[3] For a deeper understanding of how the church moved away from its roots, see "A Prophetic Calendar: The Feasts of Israel," by Jill Shannon. Shippensburg, PA; Destiny Image Publishers, 2009.

successive fulfillments throughout biblical history, until it reached a final fulfillment.

For example, in Genesis 12:10-20, during the time of famine, Abraham went to sojourn in Egypt. God judged Pharaoh for taking Abraham's wife, Sarah into his house. Later, Jacob's sons sold their younger brother, Joseph into Egyptian slavery. And during a later famine, Jacob and his family were forced to go down to Egypt, to buy grain from their long-lost brother, who was now ruler of Egypt. After 430 years, it came to pass that Abraham's descendants came out of Egypt, taking the wealth of Egypt and Joseph's bones with them. We read about their journey to the Promised Land in Exodus 12:36.

We see in this example (and in many others, too), a pattern. What happened to Abraham, happened to Jacob, and to his descendants. It also happened to a future descendant of Abraham: The Lord Jesus, as recorded in Matthew 2:16. Mary and Joseph took their son Jesus down to Egypt, during the slaughter of the baby boys in Bethlehem, as commanded by Herod (a type of Pharaoh). When it was safe to return, the Lord Jesus came out of Egypt, as a fulfillment of the prophecy of Hosea: "Out of Egypt I called My son." Yet Hosea did not prophecy about the Messiah; he was referring to the Exodus! Herein we see how the Jewish mind saw a pattern.

Matthew's gospel connects Hosea's prophecy to the young Jesus, coming up out of Egypt, and back to the land of His fathers. Although Jewish writers penned all of the New Testament, Matthew's gospel is one of the books, most specifically directed at a Jewish audience. Thus, Matthew displays this same Jewish understanding of the Old Testament: the Jewish idea that prophecy is pattern, not prediction. This "pattern" is brought into our own days, as we understand a deeper spiritual truth: We are in the final Exodus, moving to the Promised Land from this earth (typified as Egypt, which is the place of sin and corruption), to the Promised Land, where every tear will be wiped from our eyes.

A study of the epistles indicates that the apostles followed the Jewish Midrashic understanding of the Old Testament, recapitulated and patterned in the New Testament; there are undeniable patterns. For example, Paul tells the Corinthians that the experiences of the

Israelites were meant to serve as a pattern and a warning to us:

For I do not want you to be unaware, brethren, that our fathers were all under the cloud and all passed through the sea; and all were baptized into Moses in the cloud and in the sea; and all ate the same spiritual food; and all drank the same spiritual drink, for they were drinking from a spiritual rock which followed them; and the rock was Christ. Nevertheless, with most of them God was not well-pleased; for they were laid low in the wilderness. ***Now these things happened as examples for us****, so that we would not crave evil things as they also craved."* (1Cor.10:1-6).

Here, Paul is using Midrashic explanation: Christians experience an "Exodus pattern" in their journey from a life without Christ, to baptism into Christ. Again, we see this in the book of Hebrews, where the writer explains the symbolism of the Levitical priesthood and the Temple, in terms of Christ and the church:

"...who serve a copy and shadow of the heavenly things, just as Moses was warned by God when he was about to erect the tabernacle; for, 'See,' He says, 'that you make all things ***according to the pattern*** *which was shown you on the mountain'"* (Heb. 8:5).

We see the same rabbinic reasoning in Galatians 4:24 and onwards, which is the parable of the two women (Mount Sinai and Mount Zion). We also see this in Jude's epistle, which is another example of Midrashic literature.[4]

There are different kinds of prophecy in the Bible. Messianic prophecies and eschatological (end-time) prophecies are the two kinds of prophecies that are important in understanding the "last days," prior to the Lord Jesus' second coming. The Western mind, with its basis in secular humanism, born out of Greek philosophy, understands prophecy as a prediction and a fulfillment. However, to the Jewish mind of the Lord Jesus and the apostles, it was not

[4] "The Jewish People and their Sacred Scriptures in the Christian Bible." See http://www.vatican.va/roman_curia/congregations/cfaith/pcb_documents/rc_con_cfaith_doc_20020212_popolo-ebraico_en.html

a question of prediction of the prophetic word, but rather as being fulfilled, with successive fulfillments, until the final fulfillment. Prophecy was a pattern: something that is recapitulated in history. And hence, prophecy could have multiple fulfillments.

Each fulfillment of the prophetic word, like a cycle, repeats itself, and teaches us something about the ultimate fulfillment. To a first or second-century Jewish Christian, who read the Scriptures with Midrashic understanding, John's gospel, epistles and the book of Revelation, would have been seen as the pattern of the (new) Creation narrative. From Genesis to John's gospel, to John's book of Revelation, the Midrashic understanding of the "new creation" unravels with a progressive revelation, until it reaches the ultimate revelation.

- God walked the earth with man, in Genesis chapter 1.
- Both Genesis and the gospel of John commence with: *"In the beginning…"*
- God walked the ***earth*** again in John's Gospel. In Jesus, God walked the earth as the one who makes anew the creation of Genesis. The scripture readings in many traditional church services on Christmas day are readings from Genesis chapter 1 on creation, and the first chapter of John's gospel.
- In Genesis chapter 1, the Spirit of God moved on the water and brought forth Creation out of chaos.
- The same Spirit is given to Jesus' followers in John's gospel: *"But the Helper, the Holy Spirit, whom the Father will send in My name, He will teach you all things, and bring to your remembrance all that I said to you" (John 14:26).* The work of the Holy Spirit is to prepare the bride, the church, the new creation, for the marriage feast of the Lamb.
- In fact, St. Paul says the very same thing that John's gospel is saying, concerning the new creation. *"Therefore, if anyone is in Christ, he is a new creature; the old things passed away;* ***behold, new things have come****" (2 Cor. 5:17).*
- Finally, in Revelation, chapter 21, God walks again in the new, restored, and renewed earth with man, with the redeemed in Christ, His people.

"And I heard a loud voice from the throne, saying, 'Behold, the tabernacle of God is among men, and He will dwell among them, and they shall be His people, and God Himself will be among them, and He will wipe away every tear from their eyes; and there will no longer be any death; there will no longer be any mourning, or crying, or pain; the first things have passed away.' And He who sits on the throne said, ***'Behold, I am making all things new.'*** *And He said, 'Write, for these words are faithful and true.'"* (Rev. 21:3-5).

- In the book of Revelation, the Spirit and the bride say come. All of creation is now fully renewed and restored, and God walks the earth again with man.

"And I heard a loud voice from the throne, saying, "Behold, the tabernacle of God is among men." (Rev.21:3).

The fig tree, for example, in the Jewish Midrashic metaphor, represents the Tree of Life that we see in the garden in Genesis, in Ezekiel 47, and in the Book of Revelation. So when the Lord Jesus told Nathaniel, *"I saw you while you were still under the fig tree"* (John 1:48), He was not simply saying to Nathaniel that He saw him under a literal fig tree (although He did). He was telling him that He had seen him from the garden, from the beginning of creation, from the foundation of the world. We find St. Paul using the Midrashic hermeneutic by saying the same thing to the Ephesians: *"Just as He chose us in Him before the foundation of the world, that we would be holy and blameless before Him" (Eph. 1:4).*

Having this background understanding of biblical prophecy will help us understand the pattern of worship in heaven, as revealed in the book of Revelation. The Lord Jesus teaches us to pattern heaven, as He taught His disciples how they should pray: *"Our Father..., thy kingdom come, thy will be done on earth as it is* ***(patterned)*** *in heaven."* In other words, the vision given about the worship in heaven is a call to pattern our worship on earth as it is in heaven. The pattern of worship in the book of Revelation is the ultimate pattern of worship, from Genesis, all the way to Revelation.

Between Genesis and Revelation, the scriptures reveal a progressive revelation of worship. With each successive revelation of worship, there is an addition of a dimension to it.

Let me explain:

The Tabernacle of Moses

In the Exodus account, the Lord gives Moses revelation, as to how to make of the Ark of the Covenant. The throne of God, the mercy seat was covered by the wings of the two angels, the Cherubim. *"You shall make two cherubim of gold, make them of hammered work at the two ends of the mercy seat. Make one cherub at one end and one cherub at the other end; you shall make the cherubim of one piece with the mercy seat at its two ends. The cherubim shall have their wings spread upward, covering the mercy seat with their wings and facing one another; the faces of the cherubim are to be turned toward the mercy seat...* ***See that you make them after the pattern for them****, which was shown to you on the mountain" (Ex. 25:18-20,40).*

We see this pattern in two other passages in the Old Testament:

"And David arose and went with all the people who were with him to Baale-judah, to bring up from there the ark of God which is called by the Name, the very name of the Lord of hosts who is enthroned above the cherubim. (2Sam. 6:2)**."**

"*O Lord of hosts, the God of Israel, who is enthroned above the cherubim, You are the God, You alone, of all the kingdoms of the earth. You have made heaven and earth.* (Isa. 37:16)."

These scriptures refer to the throne of God, the mercy seat that was always accompanied and guarded by a high-ranking classification of mighty angelic beings, called the *Cherubim.* God appeared to Moses and to the people in His glory, which dwelt above the ark of the covenant, between the wings of the Cherubim. The "unapproachable light," God's glory, is sometimes referred to as His *"Shekinah"* glory. This word has the same Hebrew root as the word, "*shochen,*" which means, "to dwell," or, "to tabernacle." This word refers to the tangible presence of God. We see this pattern in the ultimate worship found in St. John's Revelation.

"*And I heard a loud voice from the throne, saying, behold,* ***the tabernacle of God is among men****, and He will dwell among them, and they shall be His people, and God Himself will be among them, and He will wipe away every tear from their eyes; and there will no longer be any death; there will no longer be any mourning, or crying, or pain; the first things have passed away. He who sits on the throne said, behold, I am making all things new. He said, write, for these words are faithful and true.*" (Rev.21:3-5).

The Lord desired that the earthly ark and its mercy seat be patterned after the Ark that exists in heaven. Sadly, this pattern was corrupted when Aaron abandoned the pattern of the mercy seat of God, and fashioned the golden calf. Worship shifted from the pattern that God had given to Moses, from the worship of the Creator to the worship of the creature. This is a picture of a perverted pattern. Here is a good question for the Christian church: When we worship God, are we aware that the mighty Cherubim in our midst bring down His mercy seat amidst us, and if so, for what purpose?

From the Tabernacle of Moses to the Tabernacle of David

The revelation given to David was the institution of continuous worship before the throne of God, the Ark of the Covenant. In 1Chronicles 16:4-6, David appoints the Levites to minister before the Ark of the Covenant, *continually*, with all kinds of musical instruments, with loud, crashing cymbals and with songs of praise and adoration. It was like what we now call "24/7 worship."

The tabernacle of David was a bit different from the tabernacle of Moses:

1.) David's tabernacle was more approachable to all people, at any time. In the tabernacle of Moses, only the high priest approached the mercy seat of God, and this was done only once in a year, on the Day of Atonement.

2.) There was great singing and dancing before the Lord in David's tabernacle. There was a freedom of expression in worship, with body, mind and spirit.
3.) David's worship included appointing psalmists, who composed anointed songs, which they sang to the Lord. Often these psalms expressed their prayers of intercession to the Lord.
4.) David's tabernacle was a mobile unit, which they carried physically into a war situation. This ensured God's presence and guidance with David, and gave him victory.
5.) This gives us the picture of worship, warfare and intercession intertwined as one complete whole and not a separated ministry.

From the Tabernacle of David to Worship as it is in Heaven

In heaven, John experiences all of the worship expressions, from Moses' tabernacle, to David's tabernacle, and more. The Lord Jesus taught His disciples to pray for the kingdom of God to come on earth, as it is in heaven. This is the ultimate purpose of worship, in heaven. This meant that heaven and earth could worship together, with one sound.

1.) In the book of Revelation, John is invited to join the angels and saints in heavenly worship. In other words, all of creation in heaven and on earth (creation includes the visible and the invisible) stands before God and worships Him, with all of their being.
2.) It includes seven aspects of worship, which I discuss later in this chapter.
3.) It includes intercession and warfare in worship, of which David's tabernacle was a shadow of what was to come. David's tabernacle was a mobile unit, and David took it with him when he went to war. The presence of the Lord gave David strategy for victory, and increased the boundaries of

the kingdom of God. In Revelation, at the heart of worship in heaven is warfare and intercession for the restoration and renewal of the earth. In other words, worship has a purpose in heaven and beckons the church to join in that purpose.

"*When He had taken the book, the four living creatures and the twenty-four elders fell down before the Lamb, each one holding a harp and golden bowls full of incense, which are the prayers of the saints.*" (Rev. 5:8).

What Revelation presents us with is by worship, followed by warfare. This continues unfolding, until the entire restoration and the renewal of the earth is completed. When the establishment of the kingdom of God is completed on the earth, all things are now prepared for the consummation of the ages: The "marriage feast of the Lamb," with His purified church, the spotless bride of Christ.

"Then the seventh angel sounded; and there were loud voices in heaven, saying, 'The kingdom of the world has become the kingdom of our Lord and of His Christ; and He will reign forever and ever.'" (Rev. 11:15).

"*Then I saw a new heaven and a new earth; for the first heaven and the first earth passed away, and there is no longer any sea. And I saw the holy city, new Jerusalem, coming down out of heaven from God, made ready as a bride adorned for her husband. And I heard a loud voice from the throne, saying, "Behold, the tabernacle of God is among men, and He will dwell among them, and they shall be His people, and God Himself will be among them*" (Rev. 21:1-3).

Thus, we see through the history of God's people, the Midrashic understanding of the development of worship, patterned as it is in heaven. This is a very important understanding for the present times in which we live. If our worship is to accomplish the purpose of establishing the kingdom of God on the earth, it must be patterned as it is in heaven. Our worship, followed by warfare and intercession, is a time ***to prepare the earth*** for the return of King Jesus.

I realize that many parts of the body of Christ might say, "Our liturgy, or our church service, or our prayer meetings, contain this pattern, and it's nothing new." I praise God if that is so. However, only the Holy Spirit can reveal if this is true, such that every time God's church meets to worship and intercede, that service is patterned after heaven, preparing the earth for the return of the Lord Jesus.

In this book, I will try to expound on the things I have seen and experienced of the worship in heaven, through the vision the Lord gave me. The goal of this book is to offer help to the body of Christ, in patterning their worship as it is in heaven. The vision relate to the book of Revelation, regarding worship and intercession, and it will be our guiding light. As the Psalmist says in Psalm 119:104-105, *"From Your precepts I get understanding; Therefore, I hate every false way. Your word is a lamp to my feet and a light to my path."*

"From the precepts of the Lord" could also mean, from the patterns in the word of God, we get an understanding of what the Lord is prophetically calling us to do, to ensure "hastening the coming of the day of God" (see 2 Peter 3:12).

I have tried to make this book as practical as possible, to help us pattern our worship as it is in heaven. I have not gone through the rudiments of explaining the nature of prophetic worship, or intercession. There are a number of books to read and learn from on these topics. Jill Shannon's has created an awesome worship CD, "***Sounds of Heaven***," which is based on this book. It sets a practical example of how to pattern the songs and sounds of worship in heaven, here on earth, in your prayer meetings or worship services at church. The goal of this book, along with Jill's CD, is to help the bride of Christ to come into this awesome experience, of joining with heaven in worship and intercession to prepare the earth for the coming of King Jesus. In fact, as you read this book and try it out yourself as an individual worshiper, as a group, or as a church, you will know the power of this heavenly pattern.

I will now look at the various elements of heavenly worship, which I saw in the vision, and will explain them.

The book of Revelation is punctuated with "sevens," meaning completeness. I am told that in Hebrew, seven (sheh'-vah) is from a root word, meaning "to be complete" or "full." From the law of first mention in the scriptures, we can see this clearly. The first time we encounter the "seventh" is in Genesis:

"*Thus the heavens and the earth were completed, and all their hosts. By the seventh day God completed His work which He had done, and He rested on the seventh day from all His work which He had done. Then God blessed the seventh day and sanctified it, because in it He rested from all His work which God had created and made.*" (Genesis 2:1-3).

God rested on the seventh day, because His work of creation was complete, entire, and perfect. Thus, seven represents this perfect completeness, as well as spiritual perfection. Patterns of sevens run through the scriptures, but more abundantly in the book of Revelation. It is instructive to do a group Bible study on this: the letters to the seven churches, the seven scenes of worship, the seven seals, seven trumpets, seven signs, the seven plagues and the seven visions that lead to the marriage feast of the Lamb.

In part I of this book, I will focus on the seven aspects of worship in heaven. Why is God focusing on the pattern of worship to be like that of heaven in these days? I believe as we pattern our worship, intercession and warfare, on the earth as it is in heaven, we will "hasten" the process of the return of the Lord to the earth. Scripture beckons us, as the bride of the Lamb, to do so. Every true bride longs for her husband to be with her. Every true church of the Lord Jesus Christ longs for His return to the earth. It is now time to arise and "fill our lamps with oil," and to go out to greet Him as He returns to earth, as the King of kings and the Lord of lords.

I believe it was always in the heart of God from the beginning of time, to prepare a dwelling place for His eternal Son to have an eternal bride; this pattern resounds in the Scriptures.

- In the book of Genesis, From Adam's rib, or side, God created Eve, his bride and helpmate. "*So the Lord God caused a deep sleep to fall upon the man, and he slept; then He took one of his ribs and closed up the flesh at that place. The Lord God fashioned into a woman the rib which He had taken from the man, and brought her to the man. The man said, 'This is now bone of my bones, and flesh of my flesh; She shall be called Woman, Because she was taken out of Man.'"* (Gen. 2:21-23).

- In the marriage feast of the Lamb, the church, the bride of Christ, is the redeemed. She is "taken" from the pierced side of Christ on the cross, washed by His blood and water, meaning baptism, the waters of regeneration, the restoration of fallen man, the new creation. This was the Lord Jesus' mission in His first coming, to prepare a bride for himself.

This is patterned in the Book of Revelation:

"Then I heard something like the voice of a great multitude and like the sound of many waters and like the sound of mighty peals of thunder, saying, Hallelujah! For the Lord our God, the Almighty, reigns. Let us rejoice and be glad and give the glory to Him, for the marriage of the Lamb has come and His bride has made herself ready. It was given to her to clothe herself in fine linen, bright and clean; for the fine linen is the righteous acts of the saints. Then He said to me, Write, 'Blessed are those who are invited to the marriage supper of the Lamb.' And He said to me, These are true words of God.'" (Rev. 19:6-9).

"*And I saw the holy city, new Jerusalem, coming down out of heaven from God, made ready as a bride adorned for her husband."* (Rev. 21:2).

"Then I saw a new heaven and a new earth; for the first heaven and the first earth passed away, and there is no longer any sea. And I saw the holy city, new Jerusalem, coming down out of heaven from God, made ready as a bride adorned for her husband. And I heard a loud voice from the throne, saying, 'Behold, the tabernacle of God is among men, and He will dwell among them, and they shall be His people, and God Himself will be among them, and He will wipe away every tear from their eyes; and there will no longer be any death; there will no longer be any mourning, or crying, or pain; the

first things have passed away.'" And He who sits on the throne said, 'Behold, I am making all things new.' And He said, 'Write, for these words are faithful and true.'" (Rev. 21:1-5).

The Spirit of God opened my understanding that worship in heaven, as depicted in the book of Revelation, had seven key aspects to it. My spirit could comprehend the distinctive choruses that weaved a tapestry of such divine sound as I experienced the worship in heaven. I call them the "songs" of the different beings in heaven. Each has a song, each has a role, and each song has a purpose in heaven. Worship in heaven has a definite purpose, a definite goal. These are the seven songs of the different groups in heaven:

1. **The song of the Cherubim**: The chorus of the four living creatures, the Cherubim: They sing the "Holy Holy Holy, *is* the Lord God, the Almighty, who was and who is and **who is to come."** In Hebrew, *"Kadosh."*
2. **The song of the Elders** around the throne of God, "Worthy are You, our Lord and our God, to receive glory and honor and power; for You created all things, and because of Your will they existed, and were created."
3. **The song of the Bride**: This is the calling or declarations by the nations, on the beauty of the Lord (the song of the bride). The Spirit and Bride say come.
4. **The song of the angels**: The New Song of heaven: "Worthy is the Lamb that was slain to receive power and riches and wisdom and might and honor and glory and blessing."
5. **The song of the tribes and nations**: The Earth synergizes with heaven in the song (the song of the Tribes and nations): "To Him who sits on the throne and to the Lamb be praise and honor and glory and power, forever and ever!"
6. **The song of creation**: The Victorious Song of the Overcomers (the "One Sound" of heaven and earth) – The Song of Moses: "Salvation belongs to our God!"
7. **The great Thanksgiving Song** the great three-fold Halleluiah!

> *"And the twenty-four elders, who sit on their thrones before God, fell on their faces and worshiped God, saying, 'We give You thanks, O Lord God, the Almighty, who are and who were, because You have taken Your great power and have begun to reign. And the nations were enraged, and Your wrath came, and the time came for the dead to be judged, and the time to reward Your bond-servants the prophets and the saints and those who fear Your name, the small and the great, and to destroy those who destroy the earth'"* (Rev. 11:16-18).

I realized that for our worship on earth to be effective in preparing the earth and ushering the kingdom of God as the Lord Jesus returns, it must be patterned on heaven's worship. In other words, it needs to have all seven elements or aspects of worship, as we see it in heaven. If we follow the pattern, we enter into the very throne room of God, in the way that exists in heaven.

I also saw that these seven aspects do not follow a chronological order, as they often do in many prayer meetings and church services I have attended in a variety of churches across the body of Christ. We are used to a format, where item one on the agenda happens first, and is done for the next five minutes or so, then comes item two... and so on and so forth. In heaven, the seven aspects of worship all happen together, in no particular chronological order. This is very dynamic and awesome. It is sometimes hard to explain. But I have begun to realize the deeper truth of Jesus' words, when He said in John 4:23-24, that the Father is seeking true worshippers who will worship Him in spirit and in truth. I believe He meant that the Father is seeking a body of worshippers who will pattern their worship as it is in heaven.

At the National Days of Prayer, 2010 (NDP) in Auckland, I gave an exhortation on this pattern of worship. After that, the entire congregation and the music team went into worship in this new way. We tried to pattern our music and worship as it is in heaven. The power of God came into the room in such a mighty way! It affected the electrical gadgets, it knocked some people off their feet just as they entered the meeting hall where the congregation was worshipping, and some experienced healing when no one laid hands on them or prayed for healing. We worshipped God for five hours, non-stop, and could have gone on through the night. Some of the people that have contacted me a week after this prayer conference, have said, "We are so 'high' in worship, that we just want to continue in the awesome presence of the Lord, even now." And this was just the beginning. Our pattern of worship was imperfect in every way that day at the NDP, but we made a start. Those who were present and experienced it want to continue in it, as it has changed their lives, their worship and intercession.

The Throne Room of God has much activity going on, which is presented to us in awesome splendor. The books of Daniel and Ezekiel, along with the Book of Revelation, share the same descriptions of the Throne Room of God. John sees the activity in the Throne Room in heaven, and this parallels Ezekiel 1, which also gives us a description of the four living creatures, a sea of glass like crystal in verse 22, a colorful throne room in verse 26, and the rainbow in verse 28. Fire is all around, similar to the pictures of lightning and thunder coming from the throne (see verse 27).

Daniel 7 also reveals the same imagery found in Revelation chapters 4 and 5. Here are some of the parallels:

- A throne set in heaven (compare Daniel 7:9 with Revelation 4:2)
- God is seated on the throne (compare Daniel 7:9 with Revelation 4:2)
- Fire before the throne (compare Daniel 7:9-10 with Revelation 4:5)
- Heavenly servants before the throne (compare Daniel 7:10 with Revelation 4:4, 6)
- Book/scroll before the throne (compare Daniel 7:10 with Revelation 5:1-5)
- The book/scroll is opened (compare Daniel 7:10 with Revelation 5:2-5, 9)
- The divine messianic figure approaches the throne to receive authority (compare Daniel 7:13-14 with Revelation 5:5-7, 12-13)

- Authority over every tribe, language, and people (compare Daniel 7:14 with Revelation 5:9)

Let us take a closer look at each of these seven aspects of worship. Let us see who is doing what and what the Word of God reveals to us about their heavenly worship. We do know that:

- God is enthroned by the worship of his saints
- God inhabits the praises of his people
- God build a tabernacle/temple (His presence) as we worship Him

This is important information each believer needs to know, in order to worship God and to move into the pattern of worship as it is in heaven.

Let us look at the seven aspects or dimensions of worship in heaven.

1. The chorus of the four living creatures: "Holy, Holy, Holy…"

The first of the seven dimensions in the heavenly worship scene.

"After these things I looked, and behold, a door standing open in heaven, and the first voice which I had heard, like the sound of a trumpet speaking with me, said, 'Come up here, and I will show you what must take place after these things.' Immediately I was in the Spirit; and behold, a throne was standing in heaven, and One sitting on the throne. And He who was sitting was like a jasper stone and a sardius in appearance; and there was a rainbow around the throne, like an emerald in appearance.

"Around the throne were twenty-four thrones; and upon the thrones I saw twenty-four elders sitting, clothed in white garments, and golden crowns on their heads. Out from the throne come flashes of lightning and sounds and peals of thunder. And there were seven lamps of fire burning before the throne, which are the seven Spirits of God; and before the throne there was something like a sea of glass,

like crystal; and in the center and around the throne, four living creatures full of eyes in front and behind. The first creature was like a lion, and the second creature like a calf, and the third creature had a face like that of a man, and the fourth creature was like a flying eagle. And the four living creatures, each one of them having six wings, are full of eyes around and within; and day and night they do not cease to say, 'Holy, Holy, Holy is the Lord God, The Almighty, who was and who is and who is to come'" (Rev. 4:1-8).

The "Four Living Creatures" are the "guardians" of the Throne of God. They accompany it wherever it goes. Ezekiel 1:24-28 has a very similar description of the throne of God. John mentions that they are four in number, which is the "earth number." For example, the four winds of the earth, and the four directions: north, south, east and west of the earth.

Revelation 7:1 says, "*After this I saw four angels standing at the four corners of the earth, holding the four winds of the earth, that no wind should blow on the earth, or on the sea, or upon any tree.*" This information tells that they are involved with the restoration and renewal processes of the earth. This aspect is very important in understanding intercession and warfare in worship. I will cover this in more detail in Part II of the book. For now, these four Living Creatures are a category of angels that are involved in the "re-genesis" or restoration and renewal of the earth to its former glory before the fall of man (see Genesis 3). As you may recall, there were prophecies given at Arthur's Pass, which said that God was doing a "new" thing in the earth. The "new" is all about the restoration and renewal of the earth to its former glory, for which all of creation is crying out – all eyes of creation are watching and waiting for heaven to move.

The shaking and the stripping of the earth have just begun. God is moving in power, as the church learns to patterns its worship as it is in heaven, it will learn to move with God in the restoration processes of the earth. The Lion of the Tribe of Judah will roar over the nations. Sin has contaminated the earth and destroyed its pristine former glory, as revealed in Genesis chapters 1 and 2. God is restoring and renewing the earth in these times, to its glory as is revealed in Revelation 21 and 22. The earth as we know it needs to be stripped and shaken of all that is not of God's doing, but was created by sinful man. Hebrews 12:26-29 assures us that God will shake and strip the earth of all that which does not reflect His glory.

The Four Living Creatures have eyes before and behind and within, which reveals their intelligence and understanding of all things past, present, and to come, and they are tireless in their service. The first time these Living Creatures are mentioned in the Bible is in Genesis 3:24, where they are called "Cherubim," but are not described. We find them placed at the entrance to the Garden of Eden to prevent the re-entrance of Adam and Eve, and to keep the way of the "Tree of Life."

At one of our intercessory prayer meetings in Auckland, my wife, Marinelle, had a vision of a headless man on a horse with a sword in his hand. He came riding in to the prayer hall and remained at the back (which is the entrance) of the prayer hall. She was praying at the entrance of the prayer hall for the group's protection as they worshipped. This evil spirit pointed to the overhead projector screen on which was projected the worship song, which the congregation was singing. What this evil spirit was signifying to her was that he wanted the worship that was given to God. This evil spirit could not enter the area where the congregation was worshipping, but remained outside. I believe the Cherubim kept it outside by their fiery swords. They guard the Holy place, and anything profane, kept outside.

In the book of Revelation, John is very precise in telling us what these mighty Cherubim do. They do not rest day or night, saying, "Holy, Holy, Holy, the Lord God Almighty, who was, and is, and is to come" (Rev. 4:8). To Moses God gave the pattern of the Tabernacle on Mount Sinai: "*According to all that I am going to show you, as the pattern of the tabernacle and the pattern of all its furniture, just so you shall construct it.*" (Ex. 25:9). He was instructed to make the "Ark of the Covenant" with two Cherubim upon it (see Ex. 25:10-22). These Cherubim were guardians of the "Mercy Seat." The tabernacle of Moses, was the first pattern of worship given to God's people. In this first pattern, we learn that as we worship God, He is enthroned by our worship, and He is enthroned on His mercy seat, to dispense His mercy to those who seek His mercy. Jesus reassures us of this in His sermon on the mount. "*Blessed are the merciful, for they shall receive mercy*" (Matt. 5:7).

As those schooled in the ministry of the prophetic would have learned, when we pattern our worship as heaven worships, the Cherubim open a portal between heaven and earth. The Cherubim come down in that place with God's mercy seat. You may have heard people prophesy, "This is holy ground" at a prayer meeting. Indeed it is, for God is enthroned on His mercy seat, set up by the Cherubim amongst His people. This is a very important understanding in the warfare and intercession that needs to follow worship. Our God is full of mercy, beyond our comprehension. As the earth goes through the stripping process, which will include famine, fires, earthquakes and the other calamities that the Lord Jesus warned us about. It will be a time of hardship, pain and suffering. **However, even through all His righteous judgments on the nations, His mercy is still available to those who repent and seek His face.**

During our intercessory prayer meetings, we try to pattern our worship as heaven worships, God's mercy comes down as He heals us. We have seen major healings take place, cancers healed, operations averted, and muscle atrophy healed. Sometimes He heals us, even if we have not asked for healing, His mercies are new each day, every morning. God's depth of love is His mercy towards us. In His mercy, He did not spare His only begotten Son, our Lord Jesus to die on the cross for you and me and for the whole world.

Songs and Sounds

At times of worship, the Spirit of God invites us to sing the "Holy, Holy, Holy" chorus. We need to declare this with singing, or just speaking it out that our God is merciful. He has come to take His place on His throne of mercy. This is an awesome point to be aware of in worship. His mercy endures forever (Psalm 136). The God of mercy and forgiveness is among us, and it gives us a great opportunity to intercede for the earth. His mercy endures; it tarries, as the saints worship God.

How can we pattern our worship after that which is before the throne of God in heaven? It is not possible in any way to *duplicate* worship in heaven, but we can only pattern our worship, just as Moses was told to *pattern* the tabernacle. Below is a simple way to do it within our physical limits, while our heart's desire is to please God.

One group of people could sing, "Holy, Holy, Holy, Lord God Almighty, who was, who is, and who is to come." Allow the Holy Spirit to be creative, as you build on this theme with other similar themes in your own words.

Let us call this "Group 1"

Another group, let us call it "Group 2," follows through, singing, "God of Mercy, God of Love. His mercies are new every morning."

Group 1 continues: "Holy, Holy, Holy,..."
Group 2 sings: "God of Mercy, God of Compassion, Your mercies are new every morning"

We now have a two-part chorus. If you listen to the first song on Jill Shannon's "Sounds of Heaven" CD, which is based on the teaching in this book, it can jump-start you into learning to sing multiple choruses. She has the singers doing exactly this, with many spaces for freely flowing in the Spirit.

Chapter nine in the book has some practical examples, drawn from the first song on the CD.

Singing these choruses permeates the environment with the holiness and mercy of God. The Cherubim clear the place spiritually, and they build the tabernacle of the Lord, the very mercy seat of God. The Lord comes to take His place, enthroned on the praises of His people, on His mercy seat. We are standing on holy ground. This is what Moses' tabernacle was supposed to remind the people of as they worshipped Yahweh. They would look at the tabernacle and know that the Cherubim have created a portal between heaven and earth, and brought God's mercy seat among His people. As the people worship Him, He is seated on His mercy seat, the God of mercy, waiting to dispense His mercy to those who call on His holy name.

"Oh, give ear, Shepherd of Israel, You who lead Joseph like a flock; ***You who are enthroned above the cherubim, shine forth****"* (Psalm 80:1).

"Yet You are holy, ***O You who are enthroned upon the praises of Israel****"* (Psalm 22:3).

*"****God will hear and answer them— Even the one who sits enthroned*** *from of old"* (Psalm 55:19).

The Lord sits enthroned on His mercy seat while the congregation sings the heavenly verses about His holiness. In heaven, this singing is endless. God's holy temple in heaven always exists. On earth, we see and pattern in an imperfect way the awesome worship in heaven.

I have experienced worship in the market place that has created a very different atmosphere than merely singing "Christian" songs. A group of musicians and singers were worshipping God and not *performing* for the crowd. This was at the Town Center where I live. As they got into worship, and were singing the chorus of the Cherubim, "Holy, Holy, Holy," I saw visible changes in the faces of the bystanders. The open portal drew the angels down to collect the harvest. People's hardness of heart began to change. More people responded to the call to give their lives to Christ, than when the group put up a good "Christian performance."

The group learned that when you worship in the heavenly pattern, the Cherubim create a portal between heaven and earth, the tabernacle of God in that place. God, sitting on His mercy seat, is ready to pour out His grace and mercy without measure (seen by faces and hearts changed). In the changed environment, people are more open to receiving Christ into their lives as Redeemer, Lord and Savior. I contacted the pastor and shared with him this teaching about worship. It has blessed the group's worship. The pastor said that it is going to be hard to go back to the old way of doing worship.

Worship patterned after the worship in heaven, is powerful. It is a great evangelistic way to draw people to the Lord's saving love and grace, without having to do any Bible-bashing!

Instrumental Interlude Music

In the vision I had of worship in heaven, there were some unusual sound effects, the likes of which I have not seen before. For example, I saw the halo of rainbow around the Lord's throne, but I also heard seven different musical instruments "painting the rainbow with sound," as the rainbow has seven colors. Another example would be the dazzling sound of trumpets that had a brilliance about it like the brilliance of light. Other sounds I heard were like lightning and thunder, and the sound of many rushing waters. Jill tries to incorporate these sound effects in a subtle way, to capture the atmosphere more effectively, but without being theatrical. Keep in mind, however, the atmosphere is one of tangible holiness and mercy, not that of a musical performance. In this CD, Jill has chosen to focus more on the music than on special effects, due to the difficulty of incorporating them without distracting from the worship experience.

"*And He who was sitting was like a jasper stone and a sardius in appearance; and there was a rainbow around the throne, like an emerald in appearance.*" (Rev. 4:3).

When patterning our worship as in heaven, the Spirit of God can help you develop such sound effects, without destroying the holiness of the worship atmosphere. Jill's "Sound of Heaven" CD has done this in a very simple way. Listen to the CD, and then read chapter four, where she explains some technical aspects of creating instrumental space, as well as some musical concepts, which the Lord Yeshua (Jesus) Himself taught her! As you develop the pattern of worship, you will begin to create these instrumental spaces so that people in the congregation, or in whatever format you choose, can use this part for warfare and intercession.

The music needs to reflect the ambiance of the throne room in heaven. We cannot match the sounds and songs of heaven. There are sounds and instruments in heaven, not seen or heard on the earth. We are dealing with pure spirits in heaven. They perform in more dimensions than we have on earth. Since man is both spirit and flesh, "born-again" man can understand spiritual matters through the Spirit of God.

"*But a natural* ***(fallen)*** *man does not accept the things of the Spirit of God, for they are foolishness to him; and he cannot understand them, because they are spiritually appraised. But he who is spiritual appraises all things."* (1 Cor2:14-15).

About these precious stones, sardius and jasper, there is a deeper meaning. The high priests in the times of Moses would enter the Holy of Holies, wearing a breastplate with twelve stones, representing the twelve tribes of Israel. The first stone was a Sardius, which is a deep red. Jacob's first-born son named Reuben, means, "Behold a Son." The twelfth stone was a Jasper, which is brilliant white, like a diamond. The twelfth son of Jacob was named Benjamin, which means "Son of my right hand."

Our songs should bring this aspect of the Throne – "Behold a Son, (the) Son of my Right Hand."

"*My beloved is dazzling (white) and ruddy (red), outstanding among ten thousand*" (Song of Songs 5:10).

All through your prayer meeting, try to preserve the tone of God's holiness and the mercy of God. He is present and available for your prayers of intercession, as we see this is what follows in the book of Revelation, Chapters 6 and onwards.

2. The Song of the Elders around the throne of God

This is the second of the seven dimensions of worship in heaven.

"And when the living creatures give glory and honor and thanks to Him who sits on the throne. And to Him who lives forever and ever. The twenty-four elders will fall down before Him who sits on the throne. And will worship Him who lives forever and ever. And will cast their crowns before the throne, saying, "Worthy are You, our Lord and our God, to receive glory and honor and power; for You created all things, and because of Your will they existed, and were created." (Rev. 4: 9-11).

The Elders, in their song before the throne of God, express their worship by:

- Prostrating themselves before God in heaven
- Adding to the chorus of the Cherubim, by their unique song, expressing their heart to God
- Interceding for the bride of the Lord Jesus Christ, His church

We learn from them to worship God with our entire being, our body, mind and spirit. Prostrating, kneeling, and worship dancing are all awesome bodily expressions of worship. The church must have this freedom of expression in their worship of God. This alerts the powers of darkness that we worship God alone, who alone is worthy. Paul says this in Ephesians 3:10: "*...so that the manifold wisdom of God might now be made known through the church to the rulers and the authorities in the heavenly places."*

We also learn from the elders that worship, warfare and intercession are part of the same dynamic. When the Lord Jesus was tempted in the wilderness, He employed, worship, warfare and intercession to let the devil know whom He worshipped.

"Jesus, full of the Holy Spirit, returned from the Jordan and was led around by the Spirit in the wilderness for forty days, being tempted by the devil. And He ate nothing during those days, and when they

had ended, He became hungry. and the devil said to Him, 'If You are the Son of God, tell this stone to become bread.' And Jesus answered him, 'It is written, man shall not live on bread alone.' And he led Him up and showed Him all the kingdoms of the world in a moment of time. and the devil said to Him, 'I will give You all this domain and its glory; for it has been handed over to me, and I give it to whomever I wish. Therefore if You worship before me, it shall all be Yours.' Jesus answered him, 'It is written, you shall worship the Lord your God and serve Him only.'" (Luke 4:1-8).

Who are the Elders?

Getting back to the worship in heaven in the Revelation passage above, many questions arise in our mind. Who are these twenty-four elders? Why are there only twenty-four? Why are they given thrones around the Throne of God? Why are they given crowns? One clue given to us is that this group of twenty-four is called the "Elders," while the previous group that John saw was called by a different classification, the "Four Living Creatures." Scripture will help us identify who they are, and this will help us understand how they worship God, helping us pattern our worship on earth.

The term "Elder," when used in Scripture, usually means a head of a tribe or city. The "City-elder" would be common in some small towns, even today. The "City Councilor" is a more commonly used word today. In Biblical times, the city elders played a more spiritual role than the modern governmental equivalent.

In 1 Chronicles 12:1-9, David divided the priests, for ministry in the temple of the Lord. The priesthood of Israel comprised of twenty-four heads of priestly families. John saw these twenty-four elders dressed in "white clothing." This is a further confirmation that they represent priests. In the New Testament, in 1Peter 2:9, all of God's people, the church, are the "royal priesthood," a "holy nation." We infer that the twenty-four elders that John saw before the throne of God are representative Elders of the people of God from both the Old Testament and the New Testament. The Elders in heaven represent Jewish people who have accepted "Yeshua Ha Mashiach" (Jesus the Messiah, the Anointed One), and the born-again Gentiles. This is a picture of the end-time church: the One New Man in Messiah, Jew and Gentile worshipping together.

*Therefore, remember that formerly you, Gentiles in the flesh...were at that time separate from Christ, excluded from the commonwealth of Israel, and strangers to the covenants of promise, having no hope and without God in the world. But now in Christ Jesus you who formerly were far off have been brought near by the blood of Christ. For He Himself is our peace, who made both groups into one and broke down the barrier of the dividing wall...**so that He might make the two into one new man**, thus establishing peace, and might reconcile them both in **one body to God through the cross**, by it having put to death the enmity"* (Eph. 2:11a, 12-14, 15b-16).

The Song of the Elders is the song of unity

The Elders are the ones who sing their song as 'one new man' in heaven. They are part of the one church in heaven. By "one church," I mean, not the fifty-thousand denominations we see in the body of Christ on earth. Can we say the church on earth is also one? In some places, doctrinally, yes, the church is one. However, the harsh reality is that it is a poor expression of the oneness of the church in heaven. The song of the elders is a song that calls the bride, the church on earth to that unity. On earth, we have yet to see the church operating as the 'one new man,' in every city and nation.

Often, Jewish believers in the Lord Yeshua worship in their own congregations. They feel that they are better understood, and their services follow their Jewish traditions and ancient biblical patterns. Sometimes they even face a quiet persecution that comes from the past history of the church.[5] As our worship develops the heavenly pattern, God will join the Jewish and Gentile church and make them one. This is the end time Midrashic meaning of Jesus' High Priestly prayer of John 17. Father ***make them one*** as You and I are one. The united church of the end times is an awesome church, the end time

[5] See Jill Shannon's books on this topic. "A Prophetic Calendar: The Feats of Israel," Shippensburg, PA; Destiny Image Publishers, 2009.

bride, and the glory of the Lord will shine on her. Isaiah saw this as he prophesied:

"*Arise, shine; for your light has come, and the glory of the Lord has risen upon you. For behold, darkness will cover the earth and deep darkness the peoples; But the Lord will rise upon you and His glory will appear upon you. Nations will come to your light, and kings to the brightness of your rising*" (Isa 60:1-3).

"*The latter glory of this house will be greater than the former,' says the Lord of hosts, and in this place I will give peace, declares the Lord of hosts*" (Haggai 2:9).

The Song of the Elders reminds the bride to make herself ready, for the Bridegroom is coming

The song of the Elders is also a song for the bride to get ready for the Bridegroom. The Lord Jesus came to the earth for these purposes:

- In His first coming, He ransomed with His precious blood a people for himself. The ransomed of the Lord is His bride.
- In His second coming, He comes to renew and restore the earth as a place where He and His bride, the church, will live together. (see John 14:1-3).

The Father longs to see her (the church) seated besides the Lord Jesus on the throne. The elders in heaven are seated on thrones and are crowned with crowns of righteousness; they are the bride in heaven. Their song calls the bride on the earth to be righteous, faithful and to persevere. The Lamb is worthy of these qualities from His bride. Paul mentions this:

"*I have fought the good fight, I have finished the course, I have kept the faith; in the future there is laid up for me* ***the crown of righteousness****, which the Lord, the righteous Judge, will award to me* *on that day; and not only to me,* ***but also to all who have loved His appearing***" (2Tim. 4:7-8).

"You have made him for a little while lower than the angels; ***You have crowned him with glory and honor,*** *and have appointed him over the works of your hands"* (Heb. 2:7).

The twenty-four elders prostrate themselves before the Throne of God and cast down their crowns. They are humbled by God's awesome generosity to mere man. It is here that we need to sing songs of thankfulness to God for His love, his mercy and at the same time, the songs of the Spirit and the Bride. These are also songs of intercession, where we sing of the bride's longing for her King to return to the earth. He will carry his bride with Him in his strong right arm.

We now have the second dimension of worship in place. We can hear the song of the elders before the Throne of God. They are songs of thanksgiving, they are songs of deliverance, they are songs of God's goodness and mercy, and they are songs of the Spirit and the Bride and of the longing for the coming of the King. These songs are actually intercession for the redemption of the people. At the same time, the four living creatures do not stop singing the "Holy, Holy, Holy." The pattern of worship in heaven now weaves three choruses:

Song of the Cherubim: Group #1 sings the Holy, Holy, Holy. Group #2 sings the variations on the mercy of God.

Song of the Elders: Group #3 sings songs of God's mercy, thanksgiving, deliverance, songs of the Spirit and of the Bride longing for the return of the King.

We also hear all creation groaning and longing for His return, and for the setting free of creation. These are songs with an intercessory flavor about them, a longing for the King to make haste, a desire to see the bride as "One New Man." The Psalmist expresses this desire so beautifully in the psalm below. There is longing for the

Bridegroom expressed by the bride and a call to come and deliver her, show His strength and power.

"A Song of Love. My heart overflows with a good theme; I address my verses to the King; my tongue is the pen of a ready writer. You are fairer than the sons of men; grace is poured upon Your lips; therefore God has blessed You forever. Gird Your sword on Your thigh, O Mighty One, In Your splendor and Your majesty!" (Psalm 45:1-3).

The Spirit of the Lord can help you weave these beautiful choruses. As you hear Jill's CD "Sounds of Heaven," you will experience the depth of these lyrics so artistically created by one who loves the Lord with all her heart and mind and strength echoed in her multiple chorus harmonies. Hear how exquisitely she weaves the various songs together. Listen to the words of the songs, the choruses, the instrumental interludes. They draw one to worship. As you pattern your worship from the blueprint of heaven, it will be filled with holy inspirational songs and sounds, drawing on the mercy of God in prayer and intercession. Let the Spirit of the Lord teach you as you work this through. It may not happen in the first time (and it may too!). As your heart wishes to pattern the worship in heaven, heaven will come down to help you! The depth of worship and intercession you achieve in a short time may surprise you. The Father is seeking you, a true worshipper, wanting to worship Him in spirit and truth. He will send all help to draw you into the Throne Room.

3. The song of the Bride:

"And they sang a new song, saying, 'Worthy are You to take the book and to break its seals; for You were slain, and purchased for God with Your blood men from every tribe and tongue and people and nation. You have made them to be a kingdom and priests to our God; and they will reign upon the earth'" (Rev. 5:9-10).

The worship in heaven is moving into a new phase. At this point, we need to understand that the bride, the redeemed of the Lord, is in heaven, glorified, and on the earth. One term used by many traditional churches is "the church in heaven." The church or bride in heaven are all those who have died in Christ, together with the angels, glorifying God in heaven. This understanding of the church in heaven comes from Paul's teaching in Ephesians:

"*For this reason I bow my knees before the Father, from whom every family in heaven and on earth derives its name*" (Eph. 3:14-15).

There is another scriptural passage, which speaks to us about the church in heaven:

"*But you have come to Mount Zion and to the city of the living God, the heavenly Jerusalem, and to myriads of angels, to the general assembly and church of the firstborn who are enrolled in heaven, and to God, the Judge of all, and to the spirits of the righteous made perfect, and to Jesus, the mediator of a new covenant, and to the sprinkled blood, which speaks better than the blood of Abel.*" (Heb. 12:22-24).

Paul speaks here of the "church of the firstborn who are enrolled in heaven." This church is surrounded by "myriads of angels."

John sees the church in heaven:

"*When He had taken the book, the four living creatures and the twenty-four elders fell down before the Lamb, each one holding a harp and golden bowls full of incense, which are the prayers of the saints. And they sang a new song, saying, 'Worthy are You to take the book and to break its seals; for You were slain, and purchased for God with Your blood men from every tribe and tongue and people and nation. You have made them to be a kingdom and priests to our God; and they will reign upon the earth.' Then I looked, and I heard the voice of many angels around the throne and the living creatures and the elders; and the number of them was myriads of myriads, and thousands of thousands*" (Rev. 5:8-11).

"After these things I looked, and behold, a great multitude which no one could count, from every nation and all tribes and peoples and tongues, standing before the throne and before the Lamb, *clothed in white robes, and palm branches were in their hands; and they cry out with a loud voice, saying, 'Salvation to our God who sits on the throne, and to the Lamb.' And all the angels were standing around the throne and around the elders and the four living creatures; and they fell on their faces before the throne and worshiped God"* (Rev. 7:9-11).

The bride in heaven is stirring up the Lord to take the scroll and break its seal. Why do they do this in heaven? "Worthy are You to take the book and to break its seals..." What is this all about? In the Lord Jesus' first coming to the earth, scriptures tell us that He "disarmed" the powers of darkness that rule the earth, by His obedience, death and resurrection. He "disarmed" the powers of darkness rule over the earth by taking the scroll – the title deed of the earth. The church in heaven is urging the Lord Jesus, who has won back the title deed of the earth, to now "take possession" physically. It is about taking possession of what belongs to Him. You may purchase a house and have the title deed, but never take possession by actual physical living in that house. To take possession is a legal term in any property matter. This is central to the understanding of why the earth and its inhabitants experience such violent tearing apart of humanity and the beauty of the physical earth. The reason is about taking possession of the earth. The enemy of God wants to continue possessing the earth that Adam handed over to him when he and Eve sinned. Moreover, the physical earth is dying and decaying because of the effects of sin. God is restoring the earth to its pristine glory of Genesis 1. The earth is experiencing the effects of this restoration as the old is stripped away, making room for the new.

The Lord Jesus said in Matt. 11:12, *"From the days of John the Baptist until now the kingdom of heaven suffers violence, and violent men take it by force."* Where can the kingdom of heaven suffer violence? Surely, it cannot experience violence in heaven! It suffers violence on the earth. The devil and his minions are in battle with heaven and with the church on earth, for the prime piece of real estate in the entire universe, the earth. This battle will get intense as you see its progression in the book of Revelation. The warfare reaches the supernatural state. There is the angelic battle with Michael the archangel and the powers of darkness. All this is for the possession of the earth. The earth is the dwelling place of God and the church.

The one church in heaven and on earth is more than just a theological argument that the enemy has used to divide the church on earth. It plays an important role in the return of the Lord Jesus Christ to earth. If the church in heaven is spurring the Lord Jesus Christ to break the seals and to take physical possession of the earth, what is the church on earth doing? Is the church on earth passionate about the return of her Bridegroom, the Lord Jesus? If the church in heaven and the church on earth is truly one unified church, then both should be passionate about the same thing.

The Father's heart is that the bride on earth and in heaven, the one bride of His Son, will have the same passion for the return of the Lord Jesus to the earth. He desires to see the bride on the earth as passionate for the return of her Bridegroom, as the church in heaven is passionate to see the Lord Jesus take the scroll and break its seals. As He breaks them and opens the scroll, He will come to possess the land physically, to restore and renew it, and then He will dwell with His bride on the earth. The church understands this as the second coming of the Lord Jesus.

The bride spurs the Bridegroom on, by extolling Him for all His lovely qualities, and urging Him on to take the book (scroll) and to break the seals. In doing so, the bride, the true church, reverses what Adam and Eve had done in the Garden of Eden. She takes the title deed of the earth back from the hands of the evil one and gives it to her Bridegroom, the Lord Jesus. In Psalm 2, the Father says to His Son, *"Ask of Me, and I will give You the nations for your inheritance."* The bride longs for her Bridegroom to receive the nations of the earth as His inheritance. Only He is worthy to take back the land, restoring and renewing the earth to its pristine glory. This reversal sets the warfare scenes in the book of Revelation, and we will cover this in Part II.

This is the background to the song of the bride. Firstly, it is one voice of the bride (or church), in heaven and on earth. They sing with one voice, beckoning the Bridegroom to break the seals to restore, renew, and to take the earth and reign. "You are worthy to break the seals and to reign on earth," is the core of their message. One can add several more lines to this beautiful chorus, based in the word of God. For example, Isaiah the prophet spoke that His government will have no end. Another scripture calls the Bridegroom, "The Prince of Peace." The Bridegroom is longing to hear the voice of His bride on earth! He hears the voice of this same bride in Heaven every moment. Does she long for Him? Will she endure for Him? As she joins her voice with the glorious bride in heaven and they become one voice. God is waiting, the Bridegroom is longing.

"*And I heard a loud voice from the throne, saying, 'Behold, the tabernacle of God is among men, and He will dwell among them, and they shall be His people, and God Himself will be among them, and He will wipe away every tear from their eyes; and there will no longer be any death; there will no longer be any mourning, or crying, or pain; the first things have passed away.' And He who sits on the throne said, 'Behold, I am making all things new.' And He said, 'Write, for these words are faithful and true'*" (Rev. 21:3-5).

We have already seen that the word "tabernacle" means the "shekinah" glory of the Lord. The bride in heaven enjoys God's "shekinah" glory, but she also desires that the bride on earth experience His Shekinah, His Manifest Presence. In Revelation 21, the oneness of God and man in covenant relationship is fulfilled as it was in Genesis before the fall of man.

In many of Jill's worship CDs, you will find that her songs are born out of that place of intimacy. She calls the church to love the Lord with all our heart and soul and mind and strength. The words of her songs are very lovely and full of deep meaning drawn from scriptures, reflecting God's earnest desire to be with us in covenant love. The songs of the bride are songs of covenant love. The Lord Jesus longs to hear the voice of His bride on earth sing these songs in unison with the bride in heaven.

These songs of covenant love spur the Lamb to takes the scroll in heaven, the title deed to the earth and break the seals. Only He is worthy to begin the process of breaking the seals of the scroll. This is where worship and warfare meet. The redeeming Lamb takes His place as the Lion of his pride, the Lion of his tribe, and the Lion of his bride. With the title deed legally in His hand, He will do battle for their dwelling place.

This is the place in worship where we move into intercession, as the Spirit of the Lord guides our intercessory prayers. I have learned that as an intercessor, I need to be available to God, whenever He needs me to join heaven in intercession. Most often, I find the Lord will awaken me at around 3 a.m. in the morning, or the night watch of prayer, midnight to 3 am, to pray for something. Last week, He woke me up to pray for New York City. Through sleepy eyes, I saw in the spirit "NYC." I spent the next two hours praying in the spirit for New York City. Three days later, He showed me why He called me to intercede for New York City. I read this headline on the internet: Abortions for 2009 in NYC was 41%! In other words out of every 100 babies conceived in the womb, 41 babies aborted. These statistics are passing a judgment:

1. The next generation in the US (and any nation where abortion is legal) is gradually being wiped out.
2. It is lawful to kill the unborn child in a mother's womb.
3. The nation as an entity stands before the Maker of the Universe, for breaking His law, "*You shall not kill.*" This law is one of His Ten Commandments, which are incumbent on any society to obey.

   I feared for this city and the impending judgment of God.

Below the South Wall frieze of the US Supreme Court building, which has the statue of Moses, among others, is engraved this sentence: "Justice, the guardian of liberty." The abortion figures for New York City suggest a nation where in reality another creed is lived: "Justice the guardian of immoral liberty."

Western or modern society, by its choice of a "free" lifestyle (their understanding of "liberty"), with laws to protect it, have moved more and more into a "Babylonian" type of society, which the book of Revelation predicts will ultimately fall.[6] Every time I think of those figures, I shudder to think of the fate of New York City and the US as a nation before the Lord. I can only cry out for His mercies. As we worship God, I pray that He will burden His church to intercede for our cities and nations. Yes, even before He visits us, the four horsemen that go before Him, bring the righteous judgments of our God upon our nations sins.

The words of Jesus, recorded in Luke's gospel and in the books of history, are true even today. As you read this scripture in prayer, substitute the word "Jerusalem" for the name of your city.

When He (Jesus) approached Jerusalem (replace with your city: Auckland, London, Paris, Beijing, Wellington, NYC, Philadelphia, Seattle, L.A., Vancouver, etc.), He saw the city and wept over it, saying, "If you had known in this day, even you, the things which make for peace! But now they have been hidden from your eyes. For the days will come upon you when your enemies will throw up a barricade against you, and surround you and hem you in on every side, and they will level you to the ground and your children within you, and they will not leave in you one stone upon another, ***because you did not recognize the time of your visitation.****"*

Jesus entered the temple and began to drive out those who were selling, saying to them, "It is written, 'and My house shall be a house of prayer,' but you have made it a robbers' den." (Luke 19:41-46).

---

[6] See Rev16:9; 18:2, 10, 21. I find that Chapter 18, verse 21 has an interesting word, "great millstone." "*Then a strong angel took up a stone like a great millstone and threw it into the sea, saying, 'So will Babylon, the great city, be thrown down with violence, and will not be found any longer.'*" The Lord Jesus said that if anyone did anything **to corrupt little children, which includes the corruption due to abortion**, it would be better for them to put a "millstone" around their neck and be thrown into the sea (Matt18:6). The city is made up of people, a society. These words of the Lord Jesus apply to all men, all societies, all cities (made up of a number of societies), and all nations, which are made up of many cities.

Did you see the connection that the Spirit of God is focusing on, through those scriptures?

- The Lord Jesus weeps, as He sees in the spirit realm, the coming fate of the city of Jerusalem.
- He next enters the temple to see if there is a man that will "stand in the gap" in prayer for the city.
- He does not find the "watchmen" posted on the ramparts of the city, no intercessors for the city. Instead, He finds the temple turned into a moneylenders' market place!

A stern warning resides here. The fate the city of Jerusalem met was also the fate of the temple. **The fate that lies in the balance of God's righteous judgment over a city or nation will be the fate of the church in that city or nation, if it is not a house of prayer.** Only worship, warfare and intercession joined to heaven's can change the outcome.

We have much going on in our worship at the same time, so we will need the Spirit of the Lord to bring harmony into the worship. The pattern of worship we have seen so far, paves the way to intense intercession (and warfare) in the worship. We will cover intercession in great depth in Part II.

4. The Song of the Angels, the New Song of Heaven – Worthy is the Lamb

"Then I looked, and I heard the voice of many angels around the throne and the living creatures and the elders; and the number of them was myriads of myriads, and thousands of thousands, saying with a loud voice, Worthy is the Lamb that was slain to receive power and riches and wisdom and might and honor and glory and blessing" (Rev. 5:11-12).

The bride on the earth and in heaven, with all the angels, is cheering the Lamb as He takes the scroll (but does not break the seals yet!). There is great jubilation in heaven as all heaven is waiting with all of creation for this great and awesome move of the Lamb.

The chorus of myriads and myriads of angels is an awesome sound, like the sound of many waters, which we read in Ezekiel 43:2, and Revelation 19:6.

When will the Lamb come and take the earth as His dwelling place with His bride? Only the Father knows. Heaven does not know. The earth does not know, and the bride will be surprised. This is the pattern seen in every traditional Jewish wedding, the element of surprise. The time of the bridegroom's arrival was a surprise. The bride and her bridal party were always to be ready.

This is the background of Jesus' parable of the ten virgins, recorded by Matthew 25:1-13.[7] It was customary for one of the groom's party to go ahead of the bridegroom, leading the way to the bride's house, and shout, "Behold, the Bridegroom comes." This is what it means in Revelation 22:17, where the SPIRIT and the bride say, *"Come."* The Spirit was given to the bride to get herself ready for the coming of the Bridegroom (John 14:16, 17). At the sounding of the *shofar*, the entire wedding processional would go through the streets of the city to the bride's house. That is why there are trumpet blasts in heaven, reminding us of the coming of King Jesus. Jesus warns his disciples to be ready for His return at any time of the day or hour. It is very interesting to see how Matthew and Luke portray the coming of the Lord Jesus.

"But of that day and hour no one knows, not even the angels of heaven, nor the Son, but the Father alone" (Matt. 24:36).

"For this reason you also must be ready; for the Son of Man is coming at an hour when you do not think He will" (Matt. 24:44).

"Be on guard, so that your hearts will not be weighted down with dissipation and drunkenness and the worries of life, and that day will not come on you suddenly like a trap; for it will come upon all those who dwell on the face of all the earth" (Luke 21:34-35).

[7] It is very significant that the Spirit of God placed this parable just after the Lord Jesus' Olivet discourse on the end of the age events.

The Lord has given us many things to look for, that will herald His coming. We call these the signs of the times. We cannot cover this topic in depth here. There are many volumes available, and many "insights" on this topic easily available at a local bookstore.

The 'ONE' sound in the worship in Heaven is *the sound of unity*. The "one sound" of unity of myriads and myriads of angels, and the sea of redeemed humanity in heaven, as they sing the one song, with one voice, is a sign of the unity that exists in heaven.

- The sound of one voice of the ***angelic unity in heaven*** around the throne of God.

  "Then I looked, and I heard the voice of many angels around the throne and the living creatures and the elders; and the number of them was myriads of myriads, and thousands of thousands, saying with a loud voice, 'Worthy is the Lamb that was slain to receive power and riches and wisdom and might and honor and glory and blessing.'" (Rev. 5:11-12).

- The sound of one voice of ***redeemed humanity in heaven*** before the throne of God

  "*After these things I looked, and behold, a great multitude which no one could count, from every nation and all tribes and peoples and tongues, standing before the throne and before the Lamb, clothed in white robes, and palm branches were in their hands; and they cry out with a loud voice, saying, 'Salvation to our God who sits on the throne, and to the Lamb.' And all the angels were standing around the throne and around the elders and the four living creatures; and they fell on their faces before the throne and worshiped God, saying, 'Amen, blessing and glory and wisdom and thanksgiving and honor and power and might, be to our God forever and ever. Amen.'"* (Rev. 7:9-12).

These scriptures describe a heavenly place unified under one Lord. They are one people around the throne of God, who have one mind; they desire the one thing that is deepest in the Father's heart: The return of King Jesus to earth, and the marriage feast of the Lamb. Heaven's worship has this as its one common purpose. Heaven has one united purpose, the second coming of the Lord Jesus to the earth and the restoration and renewal of the earth linked to His second coming.

We now come to the most vexing of issues, which has real repercussions on the proximity of the return of the Lord Jesus to earth. *The one unified sound of heaven, or the singular purpose of heaven's worship, is not yet the burning desire of the body of Christ on the earth!* There is a severe lack of emphasis on the return of the Lord Jesus, His second coming to the earth, in the ministries and in the teaching and preaching of the church on this vital aspect of our faith. I realize that I am walking on thin ice here, as every church will claim that this is part of their doctrine. Nevertheless, let us answer this question honestly: Does the church have this one thing in mind, this one purpose and goal in mind, in all that she is doing in her service and ministry today?

I have rarely heard a sermon focused on our purposeful preparation for the return of our Lord Jesus Christ. How many churches and ministries have made "Christian unity" their highest priority, realizing that this is the needed preparation for the return of the Lord Jesus? How many church services or televised teachings have brought their viewers closer to the true Christian unity, expressed through worship and intercession and warfare? Can you imagine the power of that expression of unity in worship as it is heaven expressed by the church on earth? I guess it would be hard to imagine something if one has never even desired it.

Some churches, denominations and other groups will sadly, be found in that state when the Lord returns, as He warns us in the parable of the ten virgins. The oil in the lamp of your sanctuary has gone out, as it had gone out among the seven churches of the book of Revelation. Those churches only exist as rubble today. The only sound you hear in that land today is the '*Adhan,*' the early morning (Muslim) call to prayer. Do they have a sound of unity, while we have lost ours? Have we lost the purpose, the goal of our worship, which is the return of the Lord Jesus to earth?

As we focus on the fourth dimension or aspect of worship, songs must reflect the Father's heart for the bride to make herself ready, and to focus on this unity. Every time we worship, we should desire unity of the body of Christ. Every time the church meets to pattern worship as it is heaven, then our desire should be that the church on earth be united as the church in heaven is united.

Musicians and songwriters have a great opportunity to express the Father's heart for the bride of his eternal Son to get ready, by being one with the rest of the body of Christ on earth and with the bride in heaven. The songs can be intercession, expressing our desire for oneness. It is an opportunity to weave that tapestry of song into our worship. It will be yet another strand of the heavenly choruses of worship that we offer to our King.

This song of the Angels is offering one unified sound with redeemed humanity in heaven, singing, "Worthy is the Lamb!" These are songs that have an intercessory groaning about them, a longing for the King to make haste, and a desire to see the bride as "One New Man."

At this point, as the Spirit leads, the prayer leader begins supplication in song, or using scriptures, and we pray for Christian Unity. The worship leader, in a sense, is "visualizing" events in heaven, and is directing the congregation in unity with heaven. This is the meaning of patterning the worship, warfare and intercession on earth as is in heaven. The church is uniting with heaven in their worship and becoming one sound, one voice.

All this is going to take a little learning, as most groups and churches are not used to this multi-layered singing and harmonizing, worshiping, interceding or moving into warfare.

If you have seen one of the movies that depict a Roman army unit (like "Spartacus"), their disciplined military movements as one unit, waged an awesome warfare. God is not so much concerned that we *perfect* our worship, as He desires us to *pattern* our worship as it is in heaven. If it were not so, He would not have taught His disciples to prayer *in like manner*, as we see in Matthew 6:9,10: "*Pray, then, in this way*: *'Our Father who is in heaven, Hallowed be Your name…thy will be done on earth as it is in heaven."* In other words, Jesus is saying, "Pattern your prayer in the way I pray." Jesus, who came down from heaven, knows what worship in heaven looks and sounds like, so He teaches us to pattern our worship on earth after the worship of His Father that takes place in heaven. As we follow Jesus, we will be doing the Father's will.

5. The song of the tribes and nations: Earth unites with heaven in the song of unity.

"To Him who sits on the throne and to the Lamb, be praise and honor and glory and power, forever and ever!"

"After these things I looked, and behold, a great multitude which no one could count, from every nation and all tribes and peoples and tongues, standing before the throne and before the Lamb, clothed in white robes, and palm branches were in their hands; and they cry out with a loud voice, saying, 'Salvation to our God who sits on the throne, and to the Lamb'" (Rev 7:9-10).

From every nation, tribe and tongue, the Lord has redeemed a people for Himself. In John 10:16, Jesus talks about the "other sheep." *"I have other sheep, which are not of this fold; I must bring them also, and they will hear My voice; and they will become one flock with one shepherd."* It is here that the tribes and nations intercede for those that have not yet "heard the voice of the Lord."

In the vision of worship in heaven that the Lord gave me, I saw the Lord go to crown those that were in heaven from the tribes and tongues and nations, before His throne. They said, "We are not worthy," and humbly refused the victor's crown. The said to the Lord, "Lord there are so many on the earth that we represent from our tribes and tongues and nations that still need to be set free." The Lord then to them, *"As you worship, I will set them free."* As earth becomes one with heaven in worship warfare and intercession, it paves the way for the "setting free" of the people on the earth. In the heavenly vision, the Seraphim and the Cherubim were sending out angels in their thousands, and they were returning with harvested sheaves in their hands (represented by the reaping of the harvest of the people we carried with us in worship). This was confirmed by a reading, almost immediately, from Psalm 126. ***He who goes to and fro weeping, carrying his bag of seed, shall indeed come again with a shout of joy, bringing his sheaves with him.***

The song of the tribes and nations opens opportunities to be creative in song and worship. The bride on earth is responding to the call of heaven to unity. Songwriters, musicians and singers can create songs of repentance, songs of deliverance, and songs of stirring up the earth to be one with heaven.

In larger congregations, we could have one group in the church singing, or representing the song of heaven, and calling earth to unity. Another group could represent the bride on the earth, responding to that call from heaven to unity. Then heaven and earth can synergize together, in a common chorus. However, it is more than a song in itself; it is a synergizing of heart and mind and will. The great chorus: "Salvation belongs to our God who sits on the throne, and to the Lamb" is a declaration to the powers of darkness, affirming allegiance to God alone, who is the savior of the world.

Moving into a new form and style of worship, is going to need some will and determination to change from one's usual format of worship which the group or church is accustomed to following. There is a change in the dynamics; there is a change in depth of passion. If you were used to a scenario where the music team did five or six straight songs and the congregation followed, then the pastor gave his sermon and the ushers made the collections, then changing to this heavenly pattern is going to require a total new overhaul. The new dynamic is now going to be quite different. One may start with the three or four usual songs, drawing the people from the outer court and into the holy place. Then the church begins to sing the various songs of heaven:

The song of the Cherubim,
The songs of the elders (by the elders of the church),
The song of the angels,
The song of the bride,
The song of the tribes and nations, and
The song of creation

These songs make up the "powerhouse" of worship; the results will lead to warfare and intercession as the Spirit leads the group/ church. There is a cry for salvation of the people of every nation and a cry for the return of the Lord for His bride. These are the groaning of the Spirit with prayers of intercession.

I will try to explain the new dynamic; sometimes I do feel short of words to describe heaven!

If the group/church is singing this part of the worship in two groups, you might have:

Group #1 takes the part of the song of the angels

"Father, your heart is yearning for the bride for your Son, the Lion and the Lamb of heaven. All of creation groans for the revelation of your children;
All of heaven cries out that the time has come for the reality of the longing in Your heart.
The cry of your Son, that they may be one as heaven is one with You, Father.
Send forth now Your Spirit and He shall renew the face of the earth."
(songwriters can have take over from here!)

Group #2 takes the part of the bride singing:

"Hearken to that sound of heaven, the sound calling the earth.
Arise my brother, my sister, for it is time to be one church, one bride;
Let us join with heaven as one, and call our Lord with one voice
For the One to come, my husband, the one I love must be seeking me
I hear His distant footsteps, my heart longs to be one with Him."
(Something to this effect; songwriters and musicians can be creative!)

When the group reaches a maturity with handling the songs, you can move in and out of the various songs of heaven, by different groups in the church/group, as all are worshipping and responding to the Lamb on the throne. The music team begins with two or three songs that are calling the congregation to focus on the Lord. The music leader begins to sense in the Spirit the congregation's response and oneness with the Lord.

1.) Then one part of the music team begins the songs of the Cherubim – Holy, Holy, Holy... They may continue with similar songs that are all about the holiness of God and the mercy of God, as they create a portal and bring the mercy seat of God in the midst of the congregation.
2.) Those who represent the elders of the church/group, should take up their songs of the elders – most often these will lead to intercession and warfare. The other half of the music team could help by starting the elders song on the right key.
3.) Another group can represent the angels (ones who have soprano and tenor voices) could take these songs – here again the music team can get them started on the right key.
4.) The air is building up with a tapestry of worship.
5.) One group representing the bride in heaven spurs on the Lamb to take the scroll and break its seals, and she spurs the bride on the earth to join her as one voice, urging the Lamb to take the scroll.
6.) Another group can represent the tribes and tongues and nations and they sing their song.
7.) The whole church/group together sings the song of creation. This is a thunderous sound in the church.

Into this dynamic, each representative group can enter into intercession and spiritual warfare. Another way to try this is to enter intercession as the Spirit of God leads, at the breaking of the seals by the Lamb of God. The leader can read the scriptures from the book of Revelation and the intercession can break out in the church. Intercession can be made for nations/tribes and tongues, for Israel, for just about anything the Lord is leading the group to pray about. Play Jill's CD, the "Sounds of Heaven," it will give you a very good and practical way to do this new dynamic in worship.

Let us look once more at the groups:

The song of creation sung by two groups in the church, represent heaven and earth. They will sing together, in unity during some parts, especially in relation to the Father's heart. What is basically happening is that the Lamb has broken a seal on the scroll in heaven.

This is the beginning of the redemption plan for planet earth, as we see in Luke 21:28: "*But when these things begin to take place, straighten up and lift up your heads, because your redemption is drawing near.*" Our songs should be songs of deliverance for the people, of the earth. They should also be songs that stir warfare (Part II of this book is on intercession, and deals with spiritual warfare as it is in heaven).

Psalm 68 is a good example of intercessory worship. Let's take a look at it.

"Let God arise, let His enemies be scattered, And let those who hate Him flee before Him. As smoke is driven away, so drive them away; as wax melts before the fire, So let the wicked perish before God. But let the righteous be glad; let them exult before God; Yes, let them rejoice with gladness. Sing to God, sing praises to His name; lift up a song for Him who rides through the deserts, Whose name is the Lord, and exult before Him. O God, when You went forth before Your people, When You marched through the wilderness, The earth quaked; The heavens also dropped rain at the presence of God; Sinai itself quaked at the presence of God, the God of Israel." (Psalm 68:1-8).

"***Kings of armies flee, they flee,*** *And she who remains at home will divide the spoil! When you lie down among the sheepfolds, You are like the wings of a dove covered with silver, and its pinions with glistening gold.* **When the Almighty scattered the kings there,** *It was snowing in Zalmon.*" (Psalm 68:12-14).

"***The chariots of God are myriads, thousands upon thousands;*** *The Lord is among them as at Sinai, in holiness.* ***You have ascended on high, You have led captive Your captives;*** *You have received gifts among men, Even among the rebellious also, that the Lord God may dwell there. Blessed be the Lord, who daily bears our burden, The God who is our salvation."* (Psalm 68:17-19).

"***They have seen Your procession, O God, The procession of my God, my King, into the sanctuary.*** *The singers went on, the musicians after them, In the midst of the maidens beating tambourines. Bless God in the congregations, Even the Lord, You who are of the fountain of Israel.*" (Psalm 68:24-26).

"*Sing to God, O kingdoms of the earth, Sing praises to the Lord, To Him who rides upon the highest heavens, which are from ancient times; behold, He speaks forth with His voice, a mighty voice. Ascribe strength to God; His majesty is over Israel And His strength is in the skies. O God,* ***You are awesome from Your sanctuary. The God of Israel Himself gives strength and power to the people. Blessed be God!***" (Psalm 68:32-35).

The Lord is looking for the fire in our passion for His return, for His kingdom, expressed in our worship. As He said in Luke 12:49, "*I have come to cast fire upon the earth; and how I wish it were already kindled*!" Do we see this passion in our churches today for His second coming?

6. The Song of creation: The Victorious Song of "Overcomers"

"*And they sang the song of Moses, the bond-servant of God, and the song of the Lamb, saying, 'Great and marvelous are Your works, O Lord God, the Almighty; righteous and true are Your ways,* ***King of the nations****! Who will not fear, O Lord, and glorify Your name? For You alone are holy;* ***for all the nations will come and worship before You, for your righteous acts have been revealed.'*** (Rev 15:3-4).

We are moving into that part of worship, which leads to intercession and warfare.

Prophetic acts in worship: Declaration

"King of the nations" is the declaration from those who overcome the pain and persecution in their time and age. The heavens are in continuous worship, as the battle for the earth rages. Earth must join her Lord in this battle through worship, warfare and intercession. The "victorious song" is the ultimate fulfillment of the prophetic word, concerning the victory of the Lord's people.

This is the song of the bride, the song of the church. There is something that the worldwide church must sing aloud prophetically, and must sing out by faith. For Hebrews 11:1 tells us, "Now faith is the assurance of things hoped for, a conviction of things not seen." Each time the church meets (in corporate worship) till the day of fulfillment, by faith she must declare, "Christ has died, Christ has risen, and Christ will come again as the King of the nations." This is not just a good theological statement, but it is a prophetic declaration. This must echo in every church in the nations, and in the heavens. As it goes out, it is reminding the powers of darkness and the princes of darkness in high places, that the Lord is coming to bend their knees and bow their heads, to the glory of the Father, that Jesus Christ is Lord in the earth, in the heavens and below the earth (Philippians 2:9-11).

Specific Prophetic Acts

***"Then Amalek came and fought against Israel** at Rephidim. So Moses said to Joshua, 'Choose men for us and go out, fight against Amalek. Tomorrow I will station myself on the top of the hill with the staff of God in my hand.' Joshua did as Moses told him, and fought against Amalek; and Moses, Aaron, and Hur went up to the top of the hill. So it came about when Moses held his hand up, that Israel prevailed, and when he let his hand down, Amalek prevailed. But Moses' hands were heavy. Then they took a stone and put it under him, and he sat on it; and Aaron and Hur supported his hands, one on one side and one on the other. Thus his hands were steady until the sun set. So Joshua overwhelmed Amalek and his people with the edge of the sword."* (Exodus 17:8-13).

In this scripture we see that Moses was worshiping God with hands lifted up in prayer. We see in this act, Moses determined whose side the battle would swing! *Prophetic action in intercessory prayer and prophetic song* is important in the act of worship. We may not realize that in the realm of the spirit, certain acts are part of the determining factor for the performance of the prophetic word. Take for example, Jeremiah's act of writing the prophetic word concerning Babylon on a scroll. Then he directed his scribe, Seraiah, to tie it to a stone and drop it in the middle of the Euphrates. While doing that prophetic act, he had to say the words that brought judgment to that nation.

"The message which Jeremiah the prophet commanded Seraiah the son of Neriah, the grandson of Mahseiah, when he went with Zedekiah the king of Judah to Babylon in the fourth year of his reign. (Now Seraiah was quartermaster.) So Jeremiah wrote in a single scroll all the calamity which would come upon Babylon, that is, all these words which have been written concerning Babylon. Then Jeremiah said to Seraiah, 'As soon as you come to Babylon, then see that you read all these words aloud, and say, You, O Lord, have promised concerning this place to cut it off, so that there will be nothing dwelling in it, whether man or beast, but it will be a perpetual desolation. And as soon as you finish reading this scroll, you will tie a stone to it and throw it into the middle of the Euphrates, and say, 'Just so shall Babylon sink down and not rise again because of the calamity that I am going to bring upon her; and they will become exhausted.' Thus far are the words of Jeremiah." (Jer. 51:59-64).

It is a common gesture to move one's hands while talking, giving expression to what one is saying. I am sure that every Christian expression of faith, both traditional and non-traditional, has some outward gestures and expressions of faith. This could be genuflecting or prostrating, or the raising of the hands. These are non-verbal declaration to the powers of darkness, that we are aligned with the Lord God Most High God. Our prophetic action in prayer should be Spirit-led, and as far as possible, be aligned to the scriptures. This will keep us away from the path of error.

Prior to a charismatic convention in Mumbai, I received a request if my team would pray for this convention, as it was just a week away, and there were just a few registrations. I took the intercessor's team to the site of the convention, and we began to intercede on site. After a short while of praying, I had a direction from the Holy Spirit to pick up a stone and ask the question: "How many stones do you see on this site?" One intercessor said, "Possibly twenty-five thousand." So I lifted up this stone in my hand to the Lord and prayed: "Lord may these 25,000 stones on this site become 25,000 living stones that will resound with Your praises and worship in this place next week." The person in charge of the registrations for the convention called me on the day of the opening of the convention. She had 25,000 full day registrations and over 50,000 evening only registrations! That convention closed with a peak attendance of 75,000 people. Our God cannot be out done in His mercy and generosity. Hallelujah!

At another time, I was in the capital of India, New Delhi, teaching a group of intercessors on prophetic intercession. Part of the course was to take the intercessors out into the city and pray in accordance with a prophetic word given to the group. The word that was given was from Psalm 24.

"This is the generation of those who seek Him, Who seek Your face – even Jacob. Lift up your heads, O gates, and be lifted up, O ancient doors, that the King of glory may come in! Who is the King of glory? The Lord strong and mighty, the Lord mighty in battle. Lift up your heads, O gates, and lift them up, O ancient doors, That the King of glory may come in! Who is this King of glory? The Lord of hosts, He is the King of glory." (Psalm 24:6-10).

To enact the prophetic word, we went through the old city that had many ancient gates to the city. We took palms in our hands, and as we came through the gates of the city (Delhi), we welcomed Jesus, the King of Glory, into the city. We finally reached the most important landmark gate of the city, called the "India Gate!" This gate houses the monument of the "unknown soldier," with a flame that is never extinguished. It is guarded by a contingent of India's armed forces. No one is allowed to pass through those gates!

As I crossed the chained boundary mark, a soldier came to me and asked me to go back behind the chain. I then told him that we are here to pray for this great nation and part of our praying we are commanded (he would understand that word as a soldier) to pass through these gates and pray. He asked me who had given us such a command. I took the opportunity, and told him that King Jesus had commanded us. He had the uttermost respect for that word that I had ever seen. He stood to attention and said, "Then very well, I will ask the commanders of the navy and air force, and see if they will allow you to pass through these gates. Even the President of India does not pass through India Gate, but only comes up to the monument and lays a wreath during the Republic Day parade." So off he went, and came back with the commanders of the other armed forces that guard that gate. They said they respected the command of King Jesus.

Next, all the soldiers from the army, navy and air force stationed at this command post formed a corridor and allowed our team to pass through the gates to the other side. They requested that while we prayed for the nation, that we would also pray for the men and women of the armed forces and their families – which we were happy to oblige. We did our prophetic act, and we welcomed King Jesus into India, as He and His heavenly army marched through triumphantly through India Gate and into the atmosphere of the capital city, New Delhi.

Perhaps you might be wondering, "What did all this achieve?" A few weeks later, there was one of the largest ecumenical outreaches in Delhi, organized by several (over a thousand!) churches together. Since Delhi is a stronghold of Hindu extremists, the organizers did not expect more than sixty thousand people to attend the evening rallies. They placed just sixty thousand chairs for their evening attendees. To their surprise, the evening rallies attracted more than 300,000 people. The police remarked that with such huge crowds at a "religious" function, it normally ends up with a riot, caused by left wing extremists. However, there was not one incident of violence, but orderly, well-mannered people who were caring for the city and for others. There was no dirt around the place, as everything taken care of very well. More than all of this, there were thousands of healings. The local hospital received a gift from the ecumenical outreach committee, of over 100,000 wheelchairs, as people who received their healing just left their wheelchairs on the ground behind them, as a testimony that King Jesus had healed them! The powers of darkness over the city of New Delhi were curtailed in their abilities by the obedience to the prophetic word, and through the prophetic act.

Some common prophetic actions are: raising hands in prayer and song, prostrating, facing a particular direction (North, South, East West), and marching around the prayer hall with the waving of flags and banners – like a victory procession. A less common manifestation is that on occasion, intercessors may have unusual actions, like travailing and "birthing" in prayer.

I witnessed an amazing thing, when our prayer team was in Mexico City, Mexico. I heard the sound of thousands of chariots in my spirit going to war. The sound was deafening. So I went up to the microphone and said what I could hear in the spirit. Then an amazing thing happened in the physical realm. Almost immediately, as if nature was waiting for this to be declared, there was the most awesome lightening that went around the prayer hall (which was circular and had glass windows all around the circular building). It was very spectacular, and it had us all wondering what was happening. Then we heard a roar of thunder in the skies above us that literally rattled the windowpanes of the prayer hall. The next thing we knew was that electrical power shut down for a short while for the entire city! There was a sense of awe and fear at what was happening.

What happened next was very visible (sometimes people look at me and wonder if I am all right in my mind, as I share this experience). A shaft of heavenly light lit the altar of this circular church. The musicians began to sing a prophetic song: "He comes as a cloud by day and the fire by night to guide us along the way," or something close to those words, I thought was so apt to the experience. In fact the music they played was like a war beat.

That night, the most tremendous spiritual battle began in the spirit, for the nation of Mexico, as intercessors prayed for Mexico as a nation. It probably went on non-stop for 12 to 16 hours of spiritual warfare prayer. A week after this prayer mission, I received an email from the leader of that prayer mission from the USA. She said that the nation of Mexico had held elections after the prayer mission week. After 70 years of a Communist party rule, the

Christian Democrat party won the election. The church in Mexico and Christian ministries experienced much more freedom of speech, through that election.

We experienced similar results in 1989, when a team of intercessors prayed for the unification of Germany. At the Berlin wall, we did a form of prophetic act and intercession, what is now commonly known as "the Jericho March." It was my first prophetic prayer action experience. We walked for some part alongside the Berlin wall that divided West and East Germany, declaring its ultimate fall, in response to the prophetic words given to prophetic intercessors of Germany and Europe. A week later, our local TV news station showed the breaking down of the wall and the reunification of West and East Germany.

I was very new to this form of intercession, and I received good teachings and guidance from many friends and fathers in the ministry of intercession for the nations. One of my spiritual fathers was the late Derek Prince, who has written a number of books, but the one that most influenced me was his book, "Shaping History through Prayer and Fasting." I was on a prayer mission team to South Africa with Derek Prince. We prayed for a political situation at the request of Mr. Nelson Mandela, as he sat in discussion with people whose political demands were not met. In response, they had avowed a blood bath in the nation. As we prayed the Lord brought peace into that situation, and averted this otherwise horrific event.

Pieter Bos of the Netherlands, who authored the awesome book, "The Nations Called," and Johannes Facius from Germany our international director of the International Fellowship of Intercessors, were some of the great intercessors who taught me much as I was sent out prayer teams on prayer missions to the nations. Johannes has written several books on intercession, including, "As in the days of Noah," which is a book for these last days.

In Israel, Lance Lambert and Eliyahu Ben-Haim, of Intercessors for Israel, gave us sound teaching on intercession for the Middle Eastern nations, through their vast knowledge and experience of that area. They are excellent Bible teachers and intercessors for the nation of Israel. It has been a great learning experience for me, ministering to nations on all the continents of the world with these humble servants of the Lord, yet who are awesome prayer warriors.

Prophetic action in intercession and song, when directed by the Holy Spirit, is effective and has a strategic place in the times ahead of us.

"*Finally, be strong in the Lord and in the strength of His might. Put on the full armor of God, so that you will be able to stand firm against the schemes of the devil. For our struggle is not against flesh and blood, but against the rulers, against the powers, against the world forces of this darkness, against the spiritual forces of wickedness in the heavenly places.*" (Eph. 6:10-12).

"*For though we walk in the flesh, we do not war according to the flesh, for the weapons of our warfare are not of the flesh, but divinely powerful for the destruction of fortresses. We are destroying speculations and every lofty thing raised up against the knowledge of God, and we are taking every thought captive to the obedience of Christ*" (2Cor. 10:3-5).

Now, let us return to the two groups of worshipers, which we were discussing. One of these signified heaven, and the other signified earth. These two have now come together to sing as one, which is a declaration to all flesh and to the powers of darkness on the earth, in the heavens and under the earth. They declare: "Salvation belongs to our God. To Him who sits upon the Throne, and unto the Lamb, be all praise, honor and thanksgiving, for He is worthy to take the scroll and open the seals." This song is an earth-shaking rendition, as it is the voice of heaven and earth as one unified voice. Then a deep silence fills the heavens and the earth. Into the stillness, one begins to hear the sound of the four living creatures singing. At first, we hear this faintly, but gradually, it gets progressively louder, until it reaches a crescendo: **"Holy, Holy, Holy, is the Lord God Almighty, who was, and who is, and who is to come."** The worship is getting extremely deep, and the best posture in which I usually find myself, is on my face, prostrated before my Lord and King. That is why the twenty-four elders and the tribes and nations are worshipping God with their crowns cast down, face down. All of creation (not just the earth, but the entire universe of the known and unknown) is watching Him as He comes to take His bride and put into effect the re-Genesis the earth as His eternal dwelling place. What is man that you are mindful of him? Yes, You made him a little lower than angels, but You crowned him with everlasting glory!

We have now looked in detail at the first six dimensions of worship. These aspects of worship are an introduction to the visions that reveal the most awesome spiritual warfare waged by the Lord Jesus, as He comes to conquer the earth. However, before I reach the seventh and last aspect of worship which is a faith aspect, there is something I need to introduce the reader to:

An Interesting Pattern of Worship, Warfare and Intercession in the Book of Revelation

As you may recall, I shared about the vision that the Lord gave me, about the worship in heaven. I would like to refresh our memories, as to the last part of that vision:

The Lord asked of the elders, "Who are these standing here?" He pointed to a group of people, as He asked this. The elders said, "These are those who loved not their lives, but gave it up for many others to enter into the blood of the Lamb." The Lord then went to them, to place crowns on their heads. And they bowed low, refusing the crown, saying, "Lord only You are worthy to wear the crown. *Lord there are so many on the earth that we represent from our tribes and tongues and nations that still need to be set free."* ***The Lord then to them, "As you worship, I will set them free." The Spirit of the living God then spoke to me. He said, "This is true worship – <u>worship that liberates the captives and sets them free to know God. Worship that restores the earth."</u>***

And Jesus returned to Galilee in the power of the Spirit, and news about Him spread through all the surrounding district. And He began teaching in their synagogues and was praised by all. And He came to Nazareth, where He had been brought up; and as was His custom, He entered the synagogue on the Sabbath, and stood up to read. And the book of the prophet Isaiah was handed to Him. And He opened the book and found the place where it was written, 'the Spirit of the Lord is upon me, because he anointed me to preach the gospel to the poor. He has sent me to proclaim release to the captives, and recovery of sight to the blind, to set free those who are oppressed, to proclaim the favorable year of the Lord.' And He closed the book, gave it back to the attendant and sat down; and the eyes of all in the synagogue were fixed on Him. And He began to say to them, 'Today this Scripture has been fulfilled in your hearing.'" (Luke 4:14-21).

As all the earth worships the Lord Jesus, He will renew and restore the earth and set it free from sin and its effects on creation. *"Worship that liberates the captives and sets them free... worship that restores the earth!"*

The book of Revelation, beginning with chapters four and five, are all about worship in heaven. The entire created order of the heavens is in worship of our God. Angels and redeemed mankind worship God together. Since heaven is of another dimension of the spirit realm, time does not play any role. Only God knows the answers to questions like, "when" does something happen. That is because only God determines the beginning and ending of any event, The Lord Jesus is the Alpha and the Omega, and He is the beginning and the end of all things. However, it is God's graciousness to let man know some of these events. The revelation of God's mind to man is the prophetic word given to us in the scriptures and in the gift of the Holy Spirit of prophecy.

"Surely the Lord God does nothing unless He reveals His secret counsel to His servants the prophets. A lion has roared! Who will not fear? The Lord God has spoken! Who can but prophesy?" (Amos 3:7-8).

In the case of the return of the Lord Jesus, no one on earth knows the day or the hour. We must look to the signs that the Lord Jesus said to look for, that herald His coming.

"*...so, you too,* ***when you see all these things, recognize that He is near, right at the door****. Truly I say to you, this generation will not pass away until all these things take place. Heaven and earth will pass away, but My words will not pass away. But of that day and hour no one knows, not even the angels of heaven, nor the Son, but the Father alone.*" (Matt. 24:33-36).

"*Now I have told you before it happens, so that when it happens, you may believe.*" (John 14:29).

The next chapter, chapter 6, is the consequence of the worship in chapters 4 and 5. What follows after worship is judgment. This pattern continues: we see a chapter in worship, followed by a chapter of judgment, but with ever-increasing intensity, until we reach chapter 19. Between chapters 4 and 19, we have six worship scenes and six judgment scenes. The seventh is the final fulfillment of prophecy and the consummation of worship and judgment. The renewed and restored earth is ready for the marriage feast of the Lamb and His bride, the church. She has made herself ready.

The book of Revelations worship and judgment scenes, follows the Genesis pattern of creation. On the seventh day, God had completed his work of creation and entered into the Shabbat rest. John's Book of Revelation sees the complete restoration and renewal of the earth in a six "scene" setting, and in the seventh scene, all restoration and renewal is complete and enters the Shabbat rest, the Shabbat day. Remember that John begins his revelation on the first day of the week, calling it the Lord's day.

Rev. 1:10-11: "***I was in the Spirit on the Lord's day****, and I heard behind me a loud voice like the sound of a trumpet, saying, 'Write in a book what you see...'*" Revelation comes to a consummation after six worship scenes, followed by six judgment scenes. The seventh day is the day of the wedding feast of His eternal Son and His eternal bride, the church. The Spirit of the Lord declares: *"The bride has made herself ready for this eternal communion of God and man."*

"*And Jesus sent Peter and John, saying, "Go and prepare the Passover for us, so that we may eat it.*" (Luke 22:8).

"And He said to them, 'I have earnestly desired to eat this Passover with you before I suffer; for I say to you, ***I shall never again eat it until it is fulfilled in the kingdom of God.'*** *And when He had taken a cup and given thanks, He said, 'Take this and share it among yourselves; for I say to you, I will not drink of the fruit of the vine from now on until the kingdom of God comes.' And when He had taken some bread and given thanks, He broke it and gave it to them, saying, 'This is My body which is given for you; do this in remembrance of Me.' And in the same way He took the cup after they had eaten, saying, 'This cup which is poured out for you is the new covenant in My blood.'"* (Luke 22:15-20).

The "Manna" that fed Israel during the wilderness years, comes to its fulfillment in the Shabbat meal that Jesus shares with His apostles, before He goes to die on the cross. This meal was a remembrance of His first coming. The "manna" was His flesh and blood (see John 6:50-58). This living bread would sustain His bride, the church, until He would come again in all power and glory to restore an renew the earth, the new promised land. The Lord Jesus is looking forward to that day of His second coming, referring to the marriage feast. By His own words, He declares: ***"I shall never again eat it until it is fulfilled in the kingdom of God."*** This fulfillment is yet to come, at the close of this age. The Passover was a remembrance, which will not be needed anymore, for it will be consummated in the "marriage of the Lamb." What began in the Garden of Eden in Genesis, has now been restored in the City of God, the New Jerusalem that comes down from heaven to earth. It has no temple, for the tabernacle, the "shekinah" glory, the abiding presence of God is with man.

When the church patterns her worship, warfare and intercession after heaven, she joins heaven in the worship that initiates the restoration and renewal of the earth. She is making herself ready. Every Shabbat meal should be a remembrance of the past and the proclamation of the future. As Jewish families and true Christians around the world already know, Passover is not only the eating of the bitter herbs of slavery, but it is also the partaking of the wine of joy, looking forward to the "next year in Jerusalem." Nevertheless, even as we enjoy a "now" fulfillment, the Bride of Messiah is also looking to the **greater fulfillment**, uttering the greater *prophetic declaration:* **"Next time in the 'New Jerusalem,' the city that comes down from heaven with our Lord."** There is far more than meets the eye in these simple statements made here. Worship, warfare and intercession are the core substance of the church's existence.

- She was born to worship her Lord.
- She must continue to fight the good fight of faith against the world, the flesh and the devil.
- She must pray for all mankind to come to the knowledge of the truth of the one mediator between God and man, the Lord Jesus Christ, Yeshua the Messiah.
- She must pray passionately for the coming of her Bridegroom, the Lord Jesus, and for the restoration of all creation to its pristine glory.
- She must understand what it means to reign and rule with the Lord Jesus, post-rapture (or post-*parousia*).[8]
- She must plan, prepare and build now for all of this. This ensures her "getting ready" for her bridegroom.
- She must not only have oil in her lamp to keep burning until the Lord Jesus returns, but enough oil that will light the world after the Lord returns.

In Part II of the book, I will dig a little bit deeper into the judgment chapters of the book of Revelation. It is my hope that this will help the reader with intercession and spiritual warfare for their cities and nations. The days we are living in are witnessing the unfolding of God's end-times agenda. The body of Christ should be engaged in worship and intercession that will lead up to the return of the Lord Jesus Christ on earth (the second coming). Heaven expects this of the church on earth. I do believe that the true church is doing this, to some extent. Some churches do it with insight, knowledge, wisdom, and with the help of the spiritual gifts of the Holy Spirit. Other churches might not go to that depth. The aim of this book is to help all of the body of Christ to understand that today our worship

---

[8] *Parousia* is the Greek word used for His coming, a word that carries the meaning of His permanent and continuing **presence** with us.

has a goal, an objective, and a precise purpose: the return of the Lord Jesus Christ to earth. When all of the body of Christ worships and prays with this goal in mind, it will "hasten the day of His return."

The table below is a guide to what is happening in heaven, in terms of worship, warfare and intercession, as John was drawn into heaven and given a glimpse. You will notice that judgments on the earth are a response of worship.

Revelation chapters 4 and 5	**Worship scene #1** Is of God and the Lamb, surrounded by the 4 Living Creatures and the 24 elders, the myriads of angels and redeemed humanity in heaven, as the Lamb takes the scroll to break open the seals: The Revelation of the Lion and the Lamb
Revelation chapter 6	**The first Judgment** – The wrath of the Lamb – He breaks open the first four seals. The four horsemen strip the earth of its provision for the nations, provision for mankind. Controlling the earth's wind and climate, they cause damage to the food chain systems of the earth, resulting in partial judgment.
Revelation chapter 7	**Worship scene #2** – Rev 7:17 "For the Lamb in the center of the throne will be their **shepherd**, *and will guide them to springs of the water of life*; and God will wipe every tear from their eyes." The Revelation of the Lamb who is also the Shepherd, makes provision for His bride. She will experience the stripping of the earth for its restoration and renewal. In the first of the judgments, the earth is stripped of its ability to provide for man, as the four horsemen of heaven ride the earth. This is reminiscent of the covenant promise in Psalm 23: 4-6: "*Even though I walk through the valley of the shadow*

	of death, I fear no evil, for You are with me; ***Your rod*** *and Your staff, they comfort me. You prepare a table before me in the presence of my enemies; You have anointed my head with oil; my cup overflows. Surely goodness and loving kindness will follow me all the days of my life, And I will dwell in the house of the Lord forever."* As covenant worshipers, we come to the Lord through our trials and tribulations, knowing that He cares and He will protect us through the time of judgment. Moreover, even if we are to bear the marks of persecution, He will be there with us.
Revelation chapter 8 and 9	**The Second Judgment** – The seven trumpets announce the destruction of the earth and prepares it for the many restoration and renewal processes, initiated through the breaking of the seals of the scroll in heaven by the Lamb of God.
Revelation chapter 11	**Worship scene #3 –** Depicts those from outside the church, who will begin to believe in God, and others whose hearts will faint at the stripping, who will choose to hate God. The prophetic acclamation unravels with Rev. 11:15-17: "*Then the seventh angel sounded; and there were loud voices in heaven, saying,* *'The kingdom of the world has become the kingdom of our Lord and of His Christ; and He will reign forever and ever.' And the twenty-four elders, who sit on their thrones before God, fell on their faces and worshiped God, saying, 'We give You thanks, O Lord God, the Almighty, who are and who were, because You have taken Your great power and have begun to reign.'"*

Revelation chapter 12 and 13	**The third judgment:** is about the war in heaven, and the casting out of Satan by the victorious church. The church conquered Satan by the blood of the Lamb and the testimony of their lives. The church does pay a price for victory by the blood of its martyrs. Today, more than at any time in history, the Christian church is persecuted, even in so-called "freedom of speech and religion" countries, which stretch the discrimination act to persecute Christians for living their lives according to what they believe is righteous.
Revelation chapter 14:1-5	**Worship scene #4** relates to a special group of people introduced by John earlier in Rev. 7:4: "*And I heard the number of those who were sealed, one hundred and forty-four thousands, having been sealed out of every tribe of the sons of Israel.*" Now we see them again in Rev. 14:3: "*And they sang as it were a new song before the throne and before the four living creatures and the elders. And no one could learn that song except the hundred and forty-four thousands who were redeemed from the earth.*" From the symbolism of the numbers it is *very likely* that these are "Jews for Jesus," or "Messianic Jews," or "Completed Jews," who have dedicated their lives to the Messiah Yeshua and to evangelism, teaching and bringing other Jews to completion in Him (Yeshua Ha Mashiach,) the family of the Lord Jesus.
Revelation chapter 14:6-20	**The judgment scenes are the preparatory acts prior to the final judgments** and the final harvest of the earth. Those who bear the mark of the beast will bear the wrath of the Lamb.

Revelation chapter 15	**Worship scene #5** The song of Moses and the song of the Lamb is the focus of this worship scene: We see the joyful praise, which includes dancing in childlike abandon, as David did before the Ark of the Covenant. The song of Moses is sung on the sea of glass and it celebrates the truth and justice of God – the bride is in all praise of the Lamb, her redeemer kinsman, and her husband. Heaven prepares for the final wrath of God on sinful humanity, with the shout of, "Our God reigns!"
Revelation chapter 16 to 18	**Climax of the judgment scenes:** The seven bowls of the wrath of God on the earth affect every aspect of creation, breaking the "old order of creation, corrupted by sin," which includes the overthrow of Babylon. In the symbolism of the Midrash, Babylon is the "world system." This is the center of seduction and humanism, placing man at the center of all things, for his personal gratification and gain. This system glorifies man at every moment in history. When Babylon perishes, economic chaos ripples through the world. Those who worshiped at this mountain pinned their hopes and trust on markets that were "bulls and bears." These are reminiscent of the "golden calf" worshippers that Aaron had led. This golden calf has now collapsed. Has not God already done this as a "dry run," in the recent worldwide economic collapse? This is a reflection that "all the kings' horsemen and all the kings' maids" are finding it hard to create a solution to put the "humpty-dumpty of world economy" together again. In addition, is not this economic collapse a shadow of what yet is to come, and has

	not God sent out a warning? This paves the way for the "new order of creation," which is the church, not the "new world order" of sinful man under the control of the anti-Christ. The redemption of the land, and of the earth takes place, for the dwelling place of the Lamb, and of our God will be with us, among His people.
Revelation chapter 19	**Worship scene #6 –** God is exalted with the four-fold Hallelujah! The last Hallelujah announces the marriage of the Lamb. The bride has made herself ready! The four living creatures and the 24 elders have been in worshiping all through these earth shaking events. Their triumphant song now joined by a multitude of voices – Heaven and Earth are ONE! Through their worship and intercession, the bride has made herself ready and the Lamb has taken back the earth as His dwelling place for Himself and His bride, the church.
Revelation chapters 20 and 21	**Worship scene #7 – Consummation.** **Chapter 20:** "*Then death and Hades were thrown into the lake of fire. This is the second death, the lake of fire. And if anyone's name was not found written in the book of life, he was thrown into the lake of fire*" (Rev 20:14-15). **Chapter 21:** "*Then I saw a new heaven and a new earth; for the first heaven and the first earth passed away, and there is no longer any sea. And I saw the holy city, new Jerusalem, coming down out of heaven from God, made ready as a bride adorned for her husband. And I heard a loud voice from the throne, saying, 'Behold, the tabernacle of God is among men, and He*

	will dwell among them, and they shall be His people, and God Himself will be among them, and He will wipe away every tear from their eyes; and there will no longer be any death; there will no longer be any mourning, or crying, or pain; the first things have passed away.' And He who sits on the throne said, 'Behold, I am making all things new.' And He said, 'Write, for these words are faithful and true.'" (Rev 21:1-5).

The worship in heaven now enters the most complex of situations:

- The Four Living Creatures and the twenty-four Elders are in continuous worship and intercession.
- God is on the Mercy Throne, always a heart of love and mercy to those who draw near to Him.
- The Lamb is at War, and all of the angels are ready to do battle on His behalf.
- Heaven and earth are cheering the Lamb, as He subjects the principalities and powers of darkness, and causes them to bend their knees and bow their heads.
- Heaven and earth are in intercession for the Lamb, as He strips the earth of the "old order," and makes all things new.
- Nations are falling as their idols fall.
- Great cheer arises as Babylon, the wicked system of man and demonic powers fall.
- The bride is making herself ready, and sings the song of Moses and the song of the Lamb
- The great Hallelujah resounds in heaven and on earth as the Lamb takes over to reign, and as the kingdoms of the earth become the kingdoms of our Lord and of His Christ.
- The announcement of the marriage feast of the Lamb, but: "Have you got your wedding garment ready?" This is a serious question from the Father of our Lord Jesus Christ[9]

---

[9] Read Matthew 22:1-14. This parable should be read in the context of the book of Revelation , to bring out the power of this parable. See also Jill Shannon's

Various traditional churches sing or recite the praises of God in the "The Glory Be (to God)" prayer in English, ("Gloria Patri" in Latin). Unfortunately, its meaning is so obscured or misunderstood today, that it is hardly ever recited publicly.

- "**Glory be to the Father and to the Son and to the Holy Spirit,**" (Worship of Triune Godhead)
- **"as it was in the beginning"** (*the earth in its pristine glory before sin, when God "tabernacle" with man, as in Genesis chapters 1 and 2*)
- **"is now and ever shall be, world (or, the earth) without end, Amen."** (*when God restores and renews the earth to its pristine glory and will God once again will tabernacle with man (see Rev. 21:3).*

It is this patterning of worship, warfare and intercession in heaven and on earth, which sets into motion and accelerates the coming of the King and the setting up of the kingdom of heaven on earth.

Songwriters, musicians, and worship leaders have a great opportunity to be creative in this sixth aspect of worship. The sound of war undergirded with the sound of peace, the sound of destruction undergirded with the sound of building up, the sound of sorrow undergirded with the sound of joy, and the sound of harvesting is undergirded with the sound of casting into the eternal fires. The sound of the great "Hallelujah, for the Lord God Almighty reigns," is heard after each spiritual act of warfare is waged.

8. The great thanksgiving song

We have finally come to the seventh and the last of the seven aspects of worship as it is in heaven.

"*And the twenty-four elders, who sit on their thrones before God, fell on their faces and worshiped God, saying,* **'We give You thanks, O**

---

dream on page 288-290 of her book, "Seduction of Christianity," Destiny Image Publishers, 2010. Which is a vivid end time call to mankind to review their options.

Lord God, the Almighty, who are and who were, because You have taken Your great power and have begun to reign'" (Rev. 11:16-17).

In these scriptures, thanksgiving is offered to God for reigning on the earth. Who would not want to thank God for reigning and ruling the earth with righteousness, justice, peace and joy? Who would not be grateful that His righteous reign fills every nation, every tribe, every tongue, every home, and every heart? This reminds us of the prophecy of Isaiah 9:6-7: "*For a child will be born to us, a son will be given to us; and the government will rest on His shoulders; and His name will be called Wonderful Counselor, Mighty God, Eternal Father, Prince of Peace. There will be no end to the increase of His government or of peace, on the throne of David and over his kingdom, to establish it and to uphold it with justice and righteousness from then on and forevermore. The zeal of the Lord of hosts will accomplish this.*"

The world is full of conspiracy theories of the coming "One World Government." The entire move to a "one world government" is the plan of the anti-Christ. When all governments come under the control of one "president," the anti-Christ can just take-over from this president, who is none other than his installed puppet. The one world government strategy is to preempt and thus, prevent the nations from coming under the government of our Lord Jesus Christ. Unfortunately, the anti-Christ's reign over the nations of the earth will be short lived. Jesus Christ will take it back from him by an utter defeat and the destruction of his rule.

The U.N. has the following prophetic scripture carved in stone, outside its main office, probably applying it to itself as a legal entity, and to its mandate!

"They shall beat their swords into plowshares and their spears into pruning hooks. Nation will not lift up sword against nation, neither shall they learn war any more. Isaiah."

Unfortunately, no human agency can do what only God can do. The U.N., in imputing this scripture to itself, has utterly failed to achieve anything close to those words. Since the birth of the U.N., there have been more wars, more civil unrests, and more deaths than at any other time in history! To me, these words stand in judgment over the U.N. and upon those who form its governing authority. They will become the millstone around the neck of the U.N., which is the forerunner and future puppet of the anti-Christ government. When the Lord Jesus returns to earth and sets up His government, then this evil government will be uprooted and cast into the sea of fire. On that day, we will see Isaiah's complete prophetic word fulfilled:

Now it will come about that in the last days, the mountain of the house of the Lord will be established as the chief of the mountains, and will be raised above the hills; and all the nations will stream to it. And many peoples will come and say, 'Come, let us go up to the mountain of the Lord, to the house of the God of Jacob; that He may teach us concerning His ways and that we may walk in His paths.' For the law will go forth from Zion And the word of the Lord from Jerusalem. And He will judge between the nations, and will render decisions for many peoples; and they will hammer their swords into plowshares and their spears into pruning hooks. Nation will not lift up sword against nation, and never again will they learn war." (Isaiah 2:2-4).

Then we shall hear the great Hallelujah proclaimed from all the nations, for His law and His word will bring true and everlasting peace. His reign (government) will be in righteousness, peace and justice. In other words, thanksgiving due to God is because **He reigns in our hearts and on the earth**.

Now some may argue that Jesus answered Pilate that His kingdom was not of this world. *"Jesus answered, 'My kingdom is not of this world. If My kingdom were of this world, then My servants would be fighting so that I would not be handed over to the Jews; but as it is, My kingdom is not of this realm.'"* (John 18:36).

What I understand is this:

In His first coming to earth, Jesus knew that unless He went to the cross, the defeat of the powers of darkness over the hearts of men, would not take place. This was His mission in His first coming to earth. He would let nothing come in the way of this, not even the kindness of his close friend, Peter. He knew that the church, His bride could only be born from His pierced side, and from the blood and water that poured out (meaning baptism). Adam's bride, Eve was created from Adam's side, from bone of his bone, and flesh of his flesh, meaning being of the same substance as Adam. So also, the Lord Jesus commanded that we eat His flesh and drink His blood in Holy Communion a condition to be part of Him, thus, making the church, His bride, of the same substance, participating in His divine nature, but not as gods, or extensions of gods as in pantheistic beliefs. (See 2Pet.1:4). This is what the restoration of man is all about, the restoration of the image of God in man, which was perfect, before the fall of man.

Until the Lord Jesus, took our sin and died on the cross and rose again on the third day, the way to the Father was not yet opened. The way to the Holy of Holies was closed, and the fiery swords of the Cherubim barred the way back to the Garden of Eden. Man must come to the foot of the cross and make up his mind, as to which kingdom he chooses to follow. The earth was not yet restored to its original state, after the fall of man.

In that sense, Jesus spoke the truth to Pontius Pilate; His kingdom was not yet born, neither in the hearts of man, nor in its dominion over the earth. The Kingdom of God was foreshadowed, foretold and patterned in all the biblical events in the Old Testament. Jesus confirms this truth in the next line of John's gospel: Therefore Pilate said to Him, "So You are a king?"

Jesus answered, "***You say correctly that I am a king. For this I have been born, and for this I have come into the world***, *to testify to the truth. Everyone who is of the truth hears My voice.*" (John 18:37).

It was only by His death and resurrection, and the coming of the Holy Spirit on the day of the feast of Pentecost, that the bride was born, the kingdom of God was born on the earth. Everything prior to that was a foreshadowing, a pattern, and a prophetic vision of what was to be.

What then, was the purpose of His second coming to earth?

The Christian world is divided over the meaning of "the second coming." Let us examine why the Lord is returning. Is He returning, merely to "rapture" the church in mid-air, and to take her to heaven? If this were the case, why would the Lord need to send a series of precise, limited, and increasing judgments and acts of destruction upon the earth? Since Paul tells us that the "law of sin and death" is already at work in the wicked, we know that sin will work its way out in the ultimate destruction of mankind.

But the book of Revelation gives us a very different picture of the coming judgments; here, we see that the judgments are for His own kingdom purposes, and not for the sake of destruction. The spotless Bride, those who are part of the Rapture, will come back with the Lord Jesus to the earth. I will cover this issue more thoroughly in Part II, as it is important for intercession and spiritual warfare. For now, we will need to be content to know that the restored and renewed earth is the dwelling place of the Lord Jesus, and of all who have accepted Him as both Lord and Savior. These are the Bride of Christ, the church. The Lord Jesus Christ became a man and dwelt upon the earth, to fulfill this twofold purpose in His Father's heart:

1. An eternal bride (the church) for His eternal Son, Our Lord Jesus Christ.
2. An eternal dwelling place for His Son, and His bride (the church), which is the restored and renewed earth.

This is so evident, as the earth is unique among the planets and galaxies, and it is different from any other object in space; the earth supports life. God sustains it by His declared word, and science tells us that life on earth is fragile. The scientists realize that the delicate balance of sustaining life on earth is remarkable! All of the theories of visitations to planet earth by the "Nephilim" or by aliens in unidentified flying objects (both of which are demonic visitations), just go to prove one thing: the earth is the most unique and attractive spot in the entire universe, prepared by God for His eternal purposes. It is sacred ground, but profaned by sin. God will cleanse and restore it, and He will renew it to its pristine glory for His Son, our Lord Jesus Christ and His bride, the church. Heaven is coming down to earth.

And I saw the holy city, the new Jerusalem, coming down out of heaven *from God, made ready as a bride adorned for her husband. And I heard a loud voice from the throne, saying, 'Behold, the tabernacle of God is among men, and He will dwell among them, and they shall be His people, and God Himself will be among them, and He will wipe away every tear from their eyes; and there will no longer be any death; there will no longer be any mourning, or crying, or pain; the first things have passed away.' And He who sits on the throne said,* ***'Behold, I am making all things new.'*** *And He said, 'Write, for these words are faithful and true."* (Rev. 21:2-5).

Why is Christ the only person worthy to take back the earth (break the seals of the scroll)?

Firstly, Jesus Christ completed this mission ***by his complete obedience to his Father***. We read in the gospels, especially the gospel of John, that He is the Father's replica on earth (Jesus' answer to Phillip in John chapter 14, demonstrates this). Christ's work on the cross, paved a way for us to return to the Father. He made a way for us to enter into the Holy of Holies, a way back to the Father's heart. He made us heirs to the riches of His kingdom as sons and daughters.

"He is the image of the invisible God, the firstborn of all creation. For by Him all things were created, both in the heavens and on earth, visible and invisible, whether thrones or dominions or rulers or authorities – all things have been created through Him and for Him. He is before all things, and in Him all things hold together. He is also head of the body, the church; and He is the beginning, the firstborn from the dead, so that He Himself will come to have first place in everything. For it was the Father's good pleasure for all the fullness to dwell in Him, and through Him to reconcile all things to Himself, having made peace through the blood of His cross;

through Him, I say, whether things on earth or things in heaven." (Col. 1:15-20).

Secondly, there are two aspects that we need to grasp:

- By his death on the cross, the Lord Jesus Christ *disarmed* the powers of darkness that once ruled the earth. To disarm the powers of darkness means to decimate their power, authority and control over the earth. This is what warfare is all about. You first disarm the enemy. When the enemy has no more "fire power", it breaks his ability to wage war. The conquering army takes control of the defeated nation. Therefore, the Lord alone has the right to take back the title deed to the earth, which Adam and Eve relinquished to the devil in the Garden of Eden.
- By his resurrection from the dead, the Lord Jesus conquered the enemy's ability to enslave mankind, and to destroy the earth by his corruptive ways. The Lord has completely *defeated* the powers of darkness and has therefore, taken back the ***right to the ownership, to the rule over the earth***. (This is also true of any warfare between nations. The vanquished relinquishes both the people and the land to the victor of the war.)

The Lord Jesus Christ has the right to the title deed to the earth, but He needs to take it physically. Everyone who buys property knows all too well, that having the title deed is a legal document, but one must also take physical ownership of the physical land to complete the purchase and give evidence of ownership.

In John's Revelation, he sees the breaking of the seals on the scroll, as an indication that the Lord Jesus Christ has initiated processes in heaven and on earth, by which He will take the earth as His own. This becomes the victorious song: The kingdoms of the earth have become the kingdoms of our Lord Jesus Christ (see Rev.11:15). The powers of darkness have conceded defeat at the cross; but they do not want to give up territory of the earth. The war is over, and now, individual battles for territory continue. Powers of darkness had entrenched themselves as lords over various parts of the earth or nations.

In the book of Daniel, we are taught that the "Prince of Persia" is a principality and power of darkness, who rules over the nations that were called the kingdom of Persia (see Dan. 10:11-13). These principalities and powers of darkness have polluted the earth through man's sin and depravity. As each seal on the scroll is broken in heaven, the Lord Jesus will send out His angels to begin the battles against these powers. The earth needs an "extreme makeover." Altars raised to demonic gods will be stripped of their control. This refers to all corrupt systems of worship, everything pertaining to man's way of life, belief systems, literature, media, government, and alternate life styles. God will destroy all of man's corrupt systems and detestable idols, before the earth is fit to be the dwelling place of His Eternal Son, our Lord Jesus and His bride. This is indeed a time of judgment, which will be heart-rending, painful, and fearsome. The time has come when the fiercest of battles will rage on the earth and in the heavens, for the sole purpose of dominion over the earth. The Lamb is the victor.

"I saw in the right hand of Him who sat on the throne a book written inside and on the back, sealed up with seven seals. And I saw a strong angel proclaiming with a loud voice, Who is worthy to open the book and to break its seals? And no one in heaven or on the earth or under the earth was able to open the book or to look into it. Then I began to weep greatly because no one was found worthy to open the book or to look into it; ***and one of the elders said to me, 'Stop weeping; behold, the Lion that is from the tribe of Judah, the Root of David, has overcome so as to open the book and its seven seals.'***

"And I saw between the throne (with the four living creatures) and the elders a Lamb standing, as if slain, having seven horns and seven eyes, which are the seven Spirits of God, sent out into all the earth. And He came and took the book out of the right hand of Him who sat on the throne. When He had taken the book, the four living creatures and the twenty-four elders fell down before the Lamb, each one holding a harp and golden bowls full of incense, which are the

prayers of the saints. And they sang a new song, saying, 'Worthy are You to take the book and to break its seals; for You were slain, and purchased for God with Your blood men from every tribe and tongue and people and nation. You have made them to be a kingdom and priests to our God; and they will reign upon the earth.'

"Then I looked, and I heard the voice of many angels around the throne and the living creatures and the elders; and the number of them was myriads of myriads, and thousands of thousands, saying with a loud voice, 'Worthy is the Lamb that was slain to receive power and riches and wisdom and might and honor and glory and blessing.' And every created thing which is in heaven and on the earth and under the earth and on the sea, and all things in them, I heard saying, 'To Him who sits on the throne, and to the Lamb, be blessing and honor and glory and dominion forever and ever.' And the four living creatures kept saying, 'Amen.' And the elders fell down and worshiped." (Rev. 5:1-14).

Thanksgiving Day

Why is "Thanksgiving Day" important? Paul, in his letter to Timothy, lists four disciplines of prayer: 1Timothy 2:1: *"First of all, then, I urge that entreaties and prayers, petitions* ***and thanksgivings****, be made on behalf of all men.*" A few nations have instituted a day of "Thanksgiving to God," as an important day of prayer to thank God for His great mercy and blessings towards their nations. This show of gratitude to God, by an entire nation, opens the nation to the blessings of God. In the bible, one such nation that obeyed God's prophetic message, was Nineveh. Nineveh received the blessing of peace for over a hundred and twenty years. As with Nineveh of antiquity, so it is with nations and cities today. When commercialism takes over the spiritual reality, wanton sin spreads like a cancer. God's wrath will soon catch up. It did so with Nineveh, with Sodom and Gomorrah, and with many others. God will deal with today's nations in His time, if repentance and turning to God is not forthcoming.

Western nations, in order to maintain separation of state from God, have closed down prayer in parliament, schools, courthouses, and other places of public importance. One can see the gradual decline of the nations that have done this. Any Christian who will do a serious study of spiritual warfare, will understand that at the very foundation of a nation's hierarchy of government, there are superimposed the controlling powers of the fallen angels. Daniel chapter 10 and Psalm 2 ("Why do the nations rage?") give a clear picture of this.

In Part II, I will cover more of the aspects of warfare and intercession, flowing out of heavenly worship, meant to strip the earth of sin that has contorted God's plans and purposes. We have now seen why our Lord Jesus is worthy to break the seals of the scroll, at which all of heaven bows before Him in worship. I see this as one of the very important aspect in the heavenly worship: the need to thank God and the Lamb of God, Jesus Christ, Lord Yeshua, that He is coming to take dominion of the earth, and He will make it a place that is worthy for Him to dwell with His bride, the church.

All of the worship and warfare that initiates the breaking of the seals on the scroll in heaven has just one purpose: to take back the control of the powers of darkness, and to reap the end time harvest on the earth. This includes the stripping and shaking of the earth, in order to restore and to renew the earth to its pristine glory and beauty, uncorrupted by the sin of fallen man.

"*For I consider that the sufferings of this present time are not worthy to be compared with the glory that is to be revealed to us. For the anxious longing of the creation waits eagerly for the revealing of the sons of God. For the creation was subjected to futility, not willingly, but because of Him who subjected it, in hope that the creation itself also will be set free from its slavery to corruption into the freedom of the glory of the children of God. For we know that the whole creation groans and suffers the pains of childbirth together until now.*" (Rom. 8:18-22).

Our worship is incomplete if we have not included thanksgiving as part of it. If we have been healed from sickness, we thank

God because He has cast out the dominion of darkness (represented by sickness), and ushered in the kingdom of God. Did not Jesus say this to the disciples of John the Baptist, when John asked Him if He, Jesus, was the Messiah?

"*When the men came to Him, they said, 'John the Baptist has sent us to You, to ask, "Are You the Expected One, or do we look for someone else?"' At that very time He cured many people of diseases and afflictions and evil spirits; and He gave sight to many who were blind. And He answered and said to them, 'Go and report to John what you have seen and heard: the blind receive sight, the lame walk, the lepers are cleansed, and the deaf hear, the dead are raised up, the poor have the gospel preached to them.*'" (Luke 7:20-22).

From the Lord's answer, we see that these signs were outward evidence of the Kingdom of God coming to the earth in healing, deliverance and the preaching of the gospel. That is why Jesus was concerned about the nine lepers, who did not come back to thank God (see Luke 17:12-18). They did not acknowledge the coming of the Kingdom of God on the earth, by thanking God for their healing. This is a fit warning to us to be thankful and welcome the dominion of God, the rule of God and the reign of God on the earth.

When we thank God, we give God the authority (legally) to take back the earth, even forcefully from the powers of darkness. How then should we incorporate the thanksgiving, which allows the Lord to have dominion over our nation and the nations of the world? We thank God for all that He is doing in the nation, and for what He will do through the intercessory worship, by faith. For faith is the assurance of things hoped for, the conviction of things not seen (see Heb. 11:1). Once again, this provides a great opportunity for creativity in worship songs of thanksgiving. The Psalms can be a great source of help. Take for example:

"*I will praise the name of God with a song, and will magnify Him with thanksgiving*. (Ps. 69:30).

"Enter into His gates with thanksgiving, and into his courts with praise: Give thanks unto Him, and bless his name. (Ps. 100:4).

"*Oh give thanks unto Jehovah; for He is good; for his loving kindness endureth for ever.*" (Ps. 136:1).

As we worship God with thanksgiving, we are letting Him know that we desire His leadership, His government, and His Lordship, over our locality, city and nation. He will then come in power and glory, to take his rightful dominion.

God has begun the process of stripping the nations to take back dominion over the earth. All of the prophetic words of Jesus, found in Matthew 24, Mark 13 and Luke 21, describe the processes by which the Lord's Kingdom will take back dominion of the earth. These passages are describing the same last days' sequence of events, which John saw in Revelation, chapters 6 through 21. Each process of the stripping of the earth is connected to the worship that is going up on in heaven. If we are to reign and rule with Christ, and have dominion over all the earth, this will only be achieved by worshiping God as He is worshiped in heaven. Not worship as we like it to be, prim and proper, orderly and in accordance with the will of man, but worship as it is in heaven, according to the desire of our God. Amen.

CHAPTER 3

THE MUSIC AND SOUND OF WORSHIP

As it is in Heaven

If we are to pattern our worship as it is offered by the angels and saints in heaven, we need a basic understanding of sound. I am not an expert, but I am happy to pass on what the Lord has shown me, as I asked about the various sounds I heard in the vision of worship in heaven.

There are some sounds and their effects that can only be heard or seen in the spirit of man, which has been quickened by God's Holy Spirit. For instance, how can one explain the sound of seven heavenly instruments that create the rainbow halo around the throne of God? Let us look at some simple earthly parallels, to help us understand this.

Sound is a pressure wave that has a particular frequency. A light wave can have color, based on its particular frequency. Both sound and light can be copied and transmitted with electro-magnetic frequencies. In the case of television, we have coded information carried on frequencies, which is used to produce colors, images and sounds.

On the other hand, using a different example, we see images displayed on our computer screen, which may be affected by music that is played electronically. Perhaps some of you have played song files on Windows Media Player. If so, you have seen that as the music plays, ever-changing lights and colors are displayed during the song. As we observe the changing colors, we may recognize that somehow the sounds of the music are changing the visual display. Internally, the computer program "interprets" the music's electronic pattern and combines it with its color pattern to display an image that is being affected by the sound.

More simply, maybe you have seen audio speakers that produce a "light and color show" while music is being played. Once again, there is an internal "code" that interprets the sounds, and produces colors for us to see.

We have a small earthly picture of how the Lord's heavenly "codes" can take sounds coming from musical instruments, and can modulate them into a different type and frequency of waves, which then create a color when played! Seven sounds that create the Violet/Indigo/Blue/Green/Yellow/Orange/Red frequencies could create a rainbow. I feel satisfied with this much of the scientific theory. What those instruments are, I do not know. In the spirit, I could perceive their sound and could see the visual effects their sounds created: the rainbow halo around the throne! (See Rev. 4:3). One day we will have an array of instruments when played, and their sounds will create light, containing a splendid variety of colors!

Sound has the power to communicate. All spoken language is sound. Silence is the absence of sound, just as darkness is the absence of light. All levels of creatures communicate through sound: birds, animals, fish, sea creatures, and man. Music and sound are tools of communication. Yet communication is not just sound, but it is a process of sending information, which has a specific meaning. Information is packaged in a media, and it is channeled by a sender to a receiver, via a medium; sound is one medium. Sound has the power to create! Communication is complete, when the information sent through the medium, has been received and understood by the receiver.

Science can surprise you, especially when it confirms the Word of God. The sound of creation still echoes in the universe, which is still being created and expanded! Scientists are straining their ears to hear it! On March 22, 2005, National Geographic reported on its web site, an interesting article on the primordial sound that created the heavens. In 1963, Nobel Laureates Arno Penzias and Robert Wilson of Bell Labs, used sophisticated scientific instruments to discover a faint microwave glow, not visible to the naked eye, across the sky. They said that this cosmic microwave background radiation represents the ancient afterglow of the "big bang." In 2001, NASA launched sophisticated microwave-imaging instruments, in a project named, "The Wilkinson Microwave Anisotropy Probe (WMAP)." It mapped the microwave spectrum in the universe. "This sound spectrum spans about ten octaves. ***The top five octaves correspond to acoustic waves, the primordial sound of the universe,*** just 380,000 years into its existence"[10] (emphasis is the author's).

I am not a scientist, nor a "big bang" theory supporter. However, this article seems to indicate to me that there is a sound lingering still in space, which went forth from heaven and created the universe. That creative sound, I believe, was the voice of God. John's gospel begins with the beginning of all things:

"*In the beginning was the Word, and the Word was with God, and the Word was God. He was in the beginning with God.* ***All things came into being through Him, and apart from Him nothing came into being that has come into being.***" (John 1:1-3).

"*In the beginning God created the heavens and the earth. And the earth was waste and void; and darkness was upon the face of the deep: and the Spirit of God moved upon the face of the waters.* ***And God said, Let there be light: and there was light.***" (Gen 1:1-3).

"God said..." created by the voice of God, the entire universe came forth. What an awesome creative sound, an awesome creative language. I wonder if the scientists who have captured that first creative sound, ever questioned in their hearts, "Is that the voice of God that said, 'Let there be...'"? These learned men have observed that this creative sound still resonates in the universe, and still creates

[10] Detailed article can be read at National Geographic's website: http://news.nationalgeographic.com/news/2004/09/0920_040920_big_bang.html

anew. Quite often, we hear on the news that astronomers have discovered a new star, a new galaxy, or a new sun, in space. That sound from God is still creating the universe.

The Sound of Healing

During a Christian healing service, people experience healing in their body, mind or spirit, by the spoken word of the person praying, or by the anointed music. In all of these cases, the person praying for the sick speaks the words God has for the sickness. The sick person is restored. I have also heard people testify of receiving a healing while in worship. At our Friday intercession prayer meetings in Auckland, a man received healing, who could not raise his hands in worship, due muscle atrophy. Then suddenly, one day he could. No one spoke a healing prayer, and no one had a word of knowledge that day. We were just deeply lost in worshipping our awesome God. As we worshipped, patterned as in heaven, God's mercy came down and healed that man. He is a very quiet sort of a person. At the end of the meeting, he asked if he could testify to God's healing love. He said that for many years, he could not raise his hands in worship, due to degenerative muscle atrophy of his arms. Now, he could raise them as high as any normal person would. Wow! The room thundered with praise to God at the end of his testimony. God has blessed him with healing, and through healing of his arms, he was able to secure a good job, which he could not do for several years. God is just so awesome in what He does. He always surprises me of the depth of His kindness, gentleness and love for us.

Tuning in to Heaven's Sounds

We are accustomed to communicating by using verbal and nonverbal expressions, such as body language, sign language, paralanguage, haptic communication (the means by which people and other animals communicate via touching), chronemics,[11] eye contact,

[11] The study of the use of time in nonverbal communication. The way we perceive time, structure our time and react to time is a powerful communication tool, and helps set the stage for the communication process.

through media, i.e., pictures, graphics and sound, and writing. The spectrum of communication that we use is indeed amazing in itself!

Sound moves in waves and frequencies. There are sounds higher than the audible range of human biological hearing, called ultra high frequencies. Some animals can hear them. But the highest of all frequencies are spiritual! Since we are part biological and part spirit, we can hear sounds within the limitations of both body (or biological), and spirit. Animals can sense communication by spirits too. To be born again from above, means coming alive in the spirit to God, who is Spirit. Jesus said, "***My sheep hear My voice,*** *and I know them, and they follow Me*" (John 10:27). Jesus being God, is both Spirit and body, and He communicates in both media. His followers can hear Him speak in both media and can communicate in both media to Him as well.

In this chapter, we will look at the sounds of heaven, so that we can pattern those sounds into our worship. Thus, we will resonate with the same sound as that in heaven. We do not often use the word "pattern," when it comes to sounds or music. We could think of heaven's worship as a "tapestry," which is an artistic piece that contains many threads that are skillfully woven together. In the same way, we create a tapestry of music in a way that resembles the tapestry of sound and music in heaven. Sometimes, there are particular words that describe sounds and music, which gives the songwriter, musicians and producers inspiration for their work this helps them use their God-given creativity. There is plenty of music from Christian groups, who are already doing this. For example, words such as "brilliance," "war like," "peaceful," or "mournful," describe a tapestry or movement of sounds, music and voices, which blend to produce the desired effect. For example, consider these passages from scripture:

"*So he answered, 'Do not fear, for those who are with us are more than those who are with them.' Then Elisha prayed and said, 'O Lord, I pray, open his eyes that he may see.' And the Lord opened the servant's eyes and he saw; and behold, the mountain was full of horses and chariots of fire all around Elisha.*" (Kings 6:16-17).

If this passage were ever put into a song, it should, I would imagine, capture the sounds that it conveys. Sounds of war, commanders shouting, the "woosh" of chariots racing, metal clashing as swords hit each other, hoofs of a million horses, and the sounds of a firestorm raging. Try to imagine the army of spirits on warhorses, ready for war, the horses snorting fire and brimstone, sounds that might put the "fear of the Lord" into the spirit of mortal man.

Let's look at another example from the Exodus story:

"*Then they set out from Succoth and camped in Etham on the edge of the wilderness. The Lord was going before them in a pillar of cloud by day to lead them on the way, and in a pillar of fire by night to give them light, that they might travel by day and by night. He did not take away the pillar of cloud by day, nor the pillar of fire by night, from before the people.* (Exodus 13:20-22)."

These scriptures never fail to fill my mind with those moments of what might have actually happened. I see men and women delivered from the Pharaoh of Egypt. I still see the fear in their eyes and faces, from what had just transpired: an all-night walk across the dry seabed, journeying between the towering walls of the Red Sea, with chariots parts, dead bodies of horses, and the horsemen of Pharaoh's army, all around there is chaotic sounds of shouts and screams. Israel crosses the Red Sea, and then, with a deafening roar, the wall disappears and there is no trace of that path between the walls of water. They are now sitting on the sands of the Sinai Peninsula, with the Pillar of Fire before them, as they worship God with thanksgiving. Let your mind capture the joyful sounds of songs of deliverance and joy, and the fearful sounds of the intimidation of the enemy. Listen with your spirit ears, to the strong declarations of faith in the God of Abraham, Isaacs and Jacob. Now add to that the sound of songs, and the background sound of the pillar of fire, and the crashing of waves coming up from the Red Sea to the shore.

Here's the unforgettable story of Elijah and Elisha…

"*As they were going along and talking, behold, there appeared a chariot of fire and horses of fire which separated the two of them.*

And Elijah went up by a whirlwind to heaven. Elisha saw it and cried out, 'My father, my father, the chariots of Israel and its horsemen!' And he saw Elijah no more. Then he took hold of his own clothes and tore them in two pieces. He also took up the mantle of Elijah that fell from him and returned and stood by the bank of the Jordan. He took the mantle of Elijah that fell from him and struck the waters and said, 'Where is the Lord, the God of Elijah?' And when he also had struck the waters, they were divided here and there; and Elisha crossed over." (2Kings 2:11-14).

Here, we see an awesome picture in words with awesome sounds. Imagine this scene: a chariot of fire, the sound of horses and flames in a raging wind! Then again, the sound of the waters of the Jordan beginning to separate into half as two walls of water move to create a pathway. Music and sounds must try to capture the essence of that moment, and to resonate with the sound of that which is timeless in heaven. Haven't we heard people say, "This piece of music is "timeless"? Music that is described as "timeless" is music that transcends time.

We refer to styles of music over the course of human history, as with fashions, as part of a certain "period" in history. Most genres of music were only enjoyed by one or two generations of that era, such as "the age of the Beatles," or the "Victorian Era." Such music has its moment and is gone. But music that is "timeless" is appreciated and loved throughout the generations. For instance, classical and baroque music, such as Bach, Handel, Beethoven, Mozart or Tchaikovsky, transcend time, and these compositions are still as beloved today as when they were created by these masters in their day. It resides in the spirit of man.

This is what I heard in the worship in heaven, which has made an indelible mark in my spirit. It continues to fill my mind and spirit, even as I worship in the "quite corner" of my room. Unfortunately, I am not a musician like Jill Shannon, to be able to craft the sounds of heaven. But Jill has done it in her awesome CD, "Sound of Heaven," and she has done it well enough for all to hear, learn and pattern the sounds and songs of heaven. She has matched her sounds and songs in accordance with the timelessness of Heaven's events in the Spirit, which draws us into the "NOW" worship of God. I believe that as Christian songwriters and musicians listen, their God-given creative talents will be stirred to create their own sounds that resonate with the timelessness of the realm of the Spirit. They will draw themes from the Word of God, and will create new musical sounds from the impartation that accompanies this book and CD.

Jill shares in the next chapter, some of the technical aspects, as well as the intimate revelations, which were involved in putting together multiple choruses, such as those that are heard in heaven. When I read what she wrote, it seemed so simple. She writes very well, and her heart is so pure to worship the Lord, that it is evident that "her heart overflows with love for the Lord, as she writes her verses to her King." This chapter will release a wealth of inspiration, as well as information about composing these kinds of songs.

Today's worship should reflect what heaven is doing in the midst of us. The body of Christ must reflect what the Lord Jesus Christ is doing in the "now." **If we are not attentive to the workings the Spirit of the Lord in these last days, our worship cannot prophetically partner with Him.** We will not be in harmony with heaven, and will not be praying and prophetically declaring His intentions through our worship. Our worship would merely be an altar built for the pleasures of man, rather than an altar built for the pure offerings of holy fire unto the Lord. The Spirit of God is calling His singers and worshipers to build such an altar. Our worship of the One seated on His throne must reflect what He is doing, which will beckon us to *"Come up here, and I will show you what must take place after this"* (see Rev. 4:1).

Here are some of the startling experiences of John in Revelation:

I was in the Spirit on the Lord's day, and I heard behind me a loud voice like the sound of a trumpet. Then I turned to see the voice that was speaking with me. And having turned I saw seven golden lamp stands; and in the middle of the lamp stands I saw one like a son of man, clothed in a robe reaching to the feet, and girded across His chest with a golden sash" (Rev. 1:10-13).

"After these things I looked, and behold, a door standing open in heaven, and the first voice which I had heard, like the sound of a trumpet speaking with me, said, 'Come up here, and I will show you what must take place after these things.'" (Rev. 4:1).

In the book of Revelation, John taken up to heaven, hears a sound, a loud voice like a trumpet. (Heaven can be loud, as was the worship in David's Tabernacle. Men of war can be loud indeed, and the Tabernacle filled with the sound of trumpets, shofars, and loud, crashing cymbals!).

- He sees awesome beings around the throne of God
- He hears singing, and is invited to participate in the worship in heaven

The worship in heaven has the most awesome sounds that cannot go unnoticed. The music and sounds move with the events that are happening around the throne of God. We must remember that heaven exists in another dimension than that of earth. On earth, time is linear. Events took place in the past or taking place in the present or yet to be fulfilled in the future. In heaven, there is no time dimension. Therefore, events exist across the time spectrum at any given moment.

This is why the Midrashic understanding of scripture is important. Prophetic events are repeated as a pattern, until the event reaches its final cycle of fulfillment. John is seeing future events that will unfold in their final prophetic fulfillment. On earth, some of these events may have had one or more fulfillment, but the final fulfillment has not yet occurred. For example, we could look at the Exodus of the Israelites to the promised land. The Exodus was one fulfillment, but this same event still awaits its final consummation in heaven, as Revelation 1 indicates.

There are many books available, which interpret the book of Revelation in a variety of different ways. Some see the events in this book as having already taken place in antiquity, as if all of these prophetic events completed in history long ago.

For example, we see four horsemen in chapter 6, which are released when the Lamb breaks the first four seals on the scroll of redemption. Some people believe that these seals and their horsemen represent four of the seven church ages throughout history, taking us into the Dark Ages by the fourth one. There is some truth in this idea, but we must not limit the Lord, who is outside of time. Even though there have been past fulfillments of this prophetic event, it has not yet reached its final fulfillment. Remember that those past church ages have not yet seen the second coming of the Lord Jesus Christ. In other words, the final fulfillment has not taken place as yet.

If a Christian believes that the four horses were released at certain moments in the past, never again to be repeated, he will misunderstand many cataclysmic events of these last days, and will not be prepared for what is coming. From my visionary experiences of the worship of heaven, I believe that the releasing of these four horsemen has taken place a number of times in various regions, during various church ages, including our present age. These horsemen are even now preparing the earth for the Lord Jesus' return to earth, as these events are happening in our day and time.

Why worship can only be Patterned as in Heaven, and not Copied

The events we read in Revelation may seem to happen in some chronological linear order. On earth, which has a linear time dimension, we can only understand events in this order: past, present and future. In heaven, past, present, are future are one moment. Now, this is hard to comprehend. Here is an easy way to grasp this concept: The Christmas parade is passing through your city and on the street you live. Your home is on the bend of the street, so you watch from your balcony, or bedroom window, the first group of people in the parade. Perhaps the band passes by, and then follow the floats, and after that, the celebrities, and so on and so forth. But the TV news people in the helicopter above your home, see the entire parade – from the band to the last group of people that make up the parade – they see it all in one moment. That is why one cannot ***replicate*** the worship in heaven, as one replicates a CD or DVD. Our worship on earth can only be a ***pattern*** of the worship in heaven. Each time we worship and try to emulate the seven aspects of worship that I explain in this chapter, we pattern our worship as it takes place in heaven. We are limited on the earth and in the universe, by the dimension of time and other dimensions lost by sin. Adam lived in more dimensions than we do, prior to sin, in the Garden of Eden. Sin has corrupted so much of life on earth, that it will require a new earth, untouched by sin, to restore the original design and intentions of the Lord. He designed and created the earth for man, who was to live in loving harmony with his Maker, walking together in the garden.

When we pattern heaven, we try to comprehend the ***timelessness*** of the original. By gazing at the original, its pattern is created within us (within our limitations), and the original is right before us. That is why we say that sound is creative. It affects the things that we intercede for, as we worship, because the original exists in heaven, and they are doing the same thing as we are doing on earth. If we grasp this concept, we will understand the power of worship, and why John records in his Revelation each worship scene in heaven follows a corresponding judgment event.

Worship is creative, but it is also purposeful in Revelation. Worship moves towards a goal, and it stirs up the passions that build towards the goal. It inspires the characters who have the power to affect the events of the goal. When you attend a prayer meeting or church service where there is no passion in the people in what they are doing, you can go home depressed, wondering if the Spirit of the Lord left before you did. This is because the fire of passion in the pastor leading the service and the faithful members, has diminished from a raging fire to burned- out embers over the period of time. It needs the breath of the Holy Spirit to blow over those burned-out embers, to restore the holy fire of pure love of the Lord, and of His end-time purposes. We read in Reveltaion 2:4: "*But I have this against you, that you have left your first love.*" May the Lord not have to point His finger to His church, His bride, if coldness exists in worship. "*I know your deeds, that you are neither cold nor hot; I wish that you were cold or hot. So because you are lukewarm,* *and neither hot nor cold, I will spit you out of My mouth."* (Rev. 3:15-16).

Worship belongs to God, and therefore our worship must reflect our concern to touch the heart of God. When we sing that beautiful worship song, "I'm coming back to the heart of worship," it is a prayer to God our Father. The meaning is this: "May my heart be aligned with Your heart Father, that my worship may be true and acceptable to You." This is the heart of Christian worship. Seeking intensely the Father's heart is at the heart of worship, and anything else will lead us down the road to idolatry, even if it reflects "Christian things." The prophecies of Ezekiel warned about this so many times.

"But the Levites who went far from Me when Israel went astray, who went astray from Me after their idols, shall bear the punishment for their iniquity. Yet they shall be ministers in My sanctuary, having oversight at the gates of the house and ministering in the house; they shall slaughter the burnt offering and the sacrifice for the people, and they shall stand before them to minister to them. Because they ministered to them before their idols and became a stumbling block of iniquity to the house of Israel, therefore I have sworn against them, declares the Lord God, that they shall bear the punishment for their iniquity. And they shall not come near to Me to serve as a priest to Me, nor come near to any of My holy things, to the things that are most holy; but they will bear their shame and their abominations which they have committed." (Ezekiel 43:10-13).

Wherever a music team exists, they should try to identify the scriptures that the Holy Spirit desires, and ask Him to help them re-create that sound sequence, the sound moment and movement. Even if you only have a guitar and a tambourine, the Holy Spirit will create a pattern of sound and song, as you move by faith. Very recently, I was asked to lead the worship in a small group of about ten to fifteen people, with just a guitar. We experienced an hour and a half of the most awesome worship and intercession for the city, and for the release of young people from their bonds of various kinds of slavery.

The effects of this worship and intercession were realized the very next day. In the heart of the city, a theater building was sold and the new owners were going to convert that building into a brothel, including a brothel for homosexuals, mainly filled by young people. The entire building just crumbled to the ground, brick upon brick. Did this incident happen because we prayed? Heaven will provide that answer to the skeptical. To those of us who prayed for the city in the midst of the heavenly courts, surrounded by the four living creatures and the twenty-four elders, surrounded by saints from every tribe, tongue and nation, and by myriads of angels, the answer was obvious. Yes, it was the result of that worship and intercession, as we sought our Father's heart.

At that moment in worship, we were not a group of ten to fifteen anymore. We were a multitude with the most awesome sound of worship and intercession, patterned and joined with heaven. God could do what He wanted to do as we prayed for the city. I have led worship in many congregations, and often I hear the familiar advice that goes like this:

"Please sing just three songs of praise, then two songs that are slow, which create a worship feeling. Then let there be silence, followed by one or two (not more), prophetic words; everything up to this point should be finished in twenty minutes, as the senior pastor needs one and a half hours for his sermon, and then there are announcements and offerings to be made."

Hallelujah! God is so patient with His church, and He shows His loving kindness and mercy, even when we offer Him the shortest time of worship, and we allot more time for the pastor to speak! The Lord Jesus had the highest regard for the temple and its worship. He was so upset that the house of prayer was turned into the "den of thieves" (see Matt. 21:13, Mark 11:17, Luke 19:46). By thieves, the Lord Jesus meant that God is robbed of the lavish worship due to Him alone from His people, not necessarily people robbing the church of tithes and offerings.

Worship in Multiple Choruses

In my vision in worship in heaven, I was captivated by the sound of multiple choruses. It was like a garden of many-colored flowers. No color dominated, but all seemed to complement each other. I could focus on any one chorus, and at the same time, hear all the choruses together like holding in my hand the entire bouquet of flowers together. Yet my eye could single out the beauty of one single flower. This is what captivated me and caused me to contact Jill Shannon, when I heard her song, "Song of the Lamb." Towards the end of the song, I heard multiple choruses, flowing together as one sound. I said to myself, "Here is someone who could understand my vision of worship in heaven."

In the heavenly worship that I experienced in the vision, I heard the prelude music, which was like showers of brilliant light with peals of lightning and thunder. I heard the sounds of many waters, rushing and bubbling up, which were fresh and cool to feel, and to the touch. The burst of sound coming from seven different instruments, seemed to create the rainbow halo around the throne. Then suddenly, the multiple choruses broke forth with clarity, like the sound of crystal in a gentle breeze, with increasing intensity.

The seven choruses that different groups sang were:

1. **Chorus of the Cherubim**: The chorus of the four living creatures, the Cherubim: "Holy Holy, Holy, Lord God Almighty…"
2. **Chorus of the Elders** around the throne of God, lifting up prayers of intercession.
3. **Chorus of the Bride** in heaven: The calling or declarations of the redeemed sea of humanity in heaven, on the beauty of the Lord. These are the songs of the Spirit and Bride, calling the bride on earth to join her in heaven as one voice, one bride on earth and in heaven, beckoning the Lamb on the throne to take the scroll and break the seals.
4. **Chorus of the angels**: The New Song of heaven. "Worthy is the Lamb…"

5. **Chorus of the tribes and nations**: "To Him who sits on the throne and to the Lamb, be praise and honor and glory and power, forever and ever!"
6. **Chorus of creation**: The Victorious Song of Overcomers. "Salvation belongs to our God."
7. **The great Thanksgiving Song**
   " *...saying, We give You thanks, O Lord God, the Almighty, who are and who were, because You have taken Your great power and have begun to reign. And the nations were enraged, and Your wrath came, and the time came for the dead to be judged, and the time to reward Your bond-servants the prophets and the saints and those who fear Your name, the small and the great, and to destroy those who destroy the earth.*" (Rev. 11:17-18).

To explain how I heard seven choruses together, and heard each one clearly, I will liken it to many ripples in the water. You can see all the ripples at one time, and then you see each ripple as they come towards you. The music seemed to play as if it was written for seven choruses. You would hear them all together, and then like a single ripple, you would hear each separate chorus as it would reach you. This is a spectacular experience.

Jill has done a marvelous work of combining these seven choruses in her songs on "Sounds of Heaven," which she created expressly for this project. It would be well-worth listening to it and emulating it. Jill and I may even consider developing training seminars for churches and groups to build (pattern) the worship of heaven into your own church. Our contact details are available at the end of the book.

Small Groups: To pattern our worship music may be a little complicated if you have just one guitarist, as happened when I was with the small group of ten to fifteen people, whose worship and intercession for young people brought the brothel building down to rubble. These groups can sing songs that have these types of choruses, or just create your own tune as we did. We had the guitarist sing the songs with choruses of: "Holy, Holy, Holy Lord." Some sang other choruses, based around the four chords the guitarist was playing. I sang songs of Tribes and Tongues and Nations, and others took up other choruses. All I know is that it was an awesome experience of one and a half hours of non-stop worship and intercession, prayer, prophetic words, declarations, and intercession, all intertwined with the worship, which wove a heavenly tapestry. It must have been an awesome sound in the Spirit, for it brought down the brothel! (You may say I have an imagination, but I have seen much, after more than twenty years of intercession among many nations. We have joined in prayer with many peoples around the world, from the slums of India to the high security prisons of Mexico. We have united in prayer with different peoples from around the world – China, Japan and Asia in the East, to Canada and the USA in the West, and from Europe in the North to Australasia in the South, and not forgetting Africa in the center! Yes indeed, I have an imagination and a joy, participating with the nations as we worship our God and see mighty changes on the earth!)

Large Groups: In larger groups and churches, you can have the most awesome experience, if you are willing to move from the normal patterns that you have established, to let heaven have its way. Or more correctly, let heaven have its **sway**, and as you join the heartbeat of heaven!

Here is an example you might wish to try:

Instrumental Music: Begin with an instrumental interlude, creating the most incredible sounds you can imagine, that "speak" about the Father, who is seated on the throne. His throne emits the most awesome light that dazzles like lightning and thunder, and gem stones and rainbows and the sound of many waters.

The Seven Choruses

Some planning and practice is needed to get a church or a large group sing and blend seven different choruses. Write separate music sheets for each of the seven choruses, which can be sung separately by different sections of the group or church. For example, the sopranos' and the tenor baritones can sing the song of the Cherubim. The church leadership sing the song of the Elders. The entire congregation can sing the song of the tribes and tongues and nations. The rest of the choir/music team sings the choruses of all the angels. In this way, you have the entire church participating in the worship of heaven. Have a practice session of singing the choruses separately and then all together. Hear the CD to understand how these choruses, blend, or sound like the ripples of water, as I explained earlier. All should participate in the church, for there are no spectators, only true worshipers, whose worship is patterned after heaven. Remember that your choruses may not necessarily be one-line choruses. Depending on the musical skills of your team, you are free to create as many lines as you can handle in your chorus tapestry. Remember, the choruses on the "Holy, Holy, Holy" theme, create a portal that brings down the mercy seat of God in the midst of His people on earth. God is enthroned on the praises of His people. His throne is His mercy seat, held by the mighty Cherubim.

The Seven choruses together with Declarations, Intercessions and Thanksgiving

While these choruses are being sung, other groups can fill the air with loud declarations of the awesome character of God. Declarations like,

"Lord, You are Merciful and Just."
"Lord, You are Loving and Faithful," etc.

Obviously, in a large church setting, you cannot have microphones all over the building. If all the people are making declarations, you will have a thunderous sound in the church.

Putting it all together:

One group sings the Song of the Cherubim.
Another group sings the song of the elders or intercedes with the elders.
Still another group sings the song of the Bride.
Yet another group sings the song of the angels.

The entire church is making declarations and intercessions.

There are moments when the entire church sings one song together, like the thanksgiving song, or the bride's song of longing for the Lord to hasten His coming to earth. There can be moments of only instrumental music, with prophetic words that will guide the next moments of worship. The music may change its tone from soft and meditative, to war-like, as the Spirit of God leads the congregation.

This will develop gradually, based on how much each church has grown in offering worship that is patterned after heaven's worship. Each time that the church meets, God has a specific goal for that meeting. Worship has a purpose, and every time that the church meets to worship, it must have the same purpose as the Father has in His heart for that church. Each time we do what the Father wants us to do, it will hasten the return of the Lord Jesus to earth. Part II focuses on intercession, as in heaven, and as revealed in Revelation. Our worship will need to birth that intercession, as is being done in heaven.[12]

Part IV: Warfare in worship

This is a very prophetic event in worship, as it deals with the Lord preparing the earth for His return. The music team must be prepared to play the sounds of war, according to the way the Lord is subduing the earth, and bringing the evil principalities and powers of darkness under His dominion and authority, under His feet. One of the ways to do this is to move with the prophetic word declared in the book of Revelations, beginning with chapter 6. These are awesome scriptures, which deal with the stripping of the earth as the Lord renews it for a dwelling place.

It is good for the pastor of the church and his prophetic team, to seek the Lord as to how to move into warfare in worship. The bride

---

[12] I have mentioned on pages 58-60, how every chapter on worship in the book of Revelation is followed by the judgment of the nations. Our worship, warfare and intercession can follow that pattern, with the Holy Spirit giving some specific things to intercede for.

in captivity is calling out to her Lord to come swiftly to release her. We know that the gospel needs to reach the ends of the earth. The church must join her Lord, in releasing those parts of the earth that need to be set free by the stripping of the earth by HIS Word. It is earth moving with heaven's "timing" releasing the prophetic word. Isaiah 55:11 tells us: "*So will My word be which goes forth from My mouth; It will not return to Me empty, without accomplishing what I desire, and without succeeding in the matter for which I sent it.*" Music and sounds can help the church reach great depths (of a variety of aspects) of spiritual warfare, moving into battle using different weapons of warfare:

"*For though we walk in the flesh, we do not war according to the flesh, for the weapons of our warfare are not of the flesh, but divinely powerful for the destruction of fortresses. We are destroying speculations and every lofty thing raised up against the knowledge of God, and we are taking every thought captive to the obedience of Christ, and we are ready to punish all disobedience, whenever your obedience is complete.* (2Cor 10:3-6)."

Some of the weapons are:

- Praying in the Spirit
- Using Scripture to put the enemy to flight, as it is the Sword of the Spirit.
- Love and compassion to those who are under slavery.
- Standing in the gap in repentance, speaking words of release to kindred spirits
- Calling on God's Mercy, which is freely available, as God is seated on his Mercy Throne by the mighty Cherubim, as the worshipers continue to sing "Holy, Holy, Holy is the Lord."

Let's now summarize all that is happening in worship patterned as it is in heaven:

- You can begin the time of worship with instrumental music that is "speaking" of God's awesome Glory, the brilliance of Him who is on the Throne, Father God in dazzling array.
- His Throne is the Mercy Seat, held by the Four living creatures.
- The seven choruses continue throughout the worship time.
- There are Scriptural declarations about the character of God, and of the Lamb, and of the Lion of Judah. Songs on this could be very creative, and can be blended into the seven choruses.
- There are intercessions going up, while the choruses go on.
- Then suddenly, the music changes to warfare music. There is spiritual warfare going on for the nation/s that are being subdued and the earth being restored and renewed for the dwelling place of the Lamb and the bride, the church.
- The great and final chorus – The great thanksgiving song is a crescendo going up to God by heaven and earth, in the chorus of all created things, for Christ has made the kingdoms of the earth His kingdom. The earth is restored and renewed to its pristine glory of Genesis chapters 1 and 2. The marriage feast of the Lamb is announced and the music is fitting for this announcement.

At the very heart of all that we do in worship, is the Father's own heart. If we cannot get our beat to the beat of His heart, we are playing offbeat music! We are not preparing the earth for the return of His Son for His bride. There is a delay in the marriage feast. We do not want that to happen, do we?

Let us seek the Father's heart with all of our heart, as we worship here on earth, as is done heaven.

CHAPTER 4

CREATING THE SOUNDS OF HEAVEN

by Jill Shannon

A Songwriter's Journey into the New Sound

Introduction

This chapter is on the practical and spiritual side of creating heavenly music, and it is being woven into the other teachings in this groundbreaking new book. As a psalmist and lovesick worshiper, I am sharing my heart with my dear brothers and sisters, with all who desire to offer the highest praises and the deepest expressions of worship to our Father God and our Beloved Bridegroom, the Lord Jesus Christ.

Before I plunge into the heart of worship, could I take a few moments to introduce myself, and to explain how I came to be connected with this project? It is an awesome testimony of a divine appointment with destiny.

My name is Jill Shannon, and I have been a Jewish believer in the Lord Jesus since 1973. I'm honored and excited to be partnering with my new brother Robert Misst in this awesome, last days' project about the worship and intercessions taking place in the Throne Room of Heaven.

A Divine Appointment

How did I come to be writing this chapter? Well, Robert Misst, the author of this book, sent me an email one day. We were total strangers to each other at that time. He had purchased a download of my most recent worship CD, "Song of the Lamb," which was the fifth worship project I had produced at that time. His email was a kind and encouraging response to the depth of worship he heard on that project.

I've received lovely emails before, from precious saints whose spirits are touched by the music the Lord has given me. But Bob's email was unusual, and I read it carefully. In a humble and unassuming manner, he shared that he has been taken to the Throne Room of Heaven, and that he has been privileged to hear, experience and participate in the worship and intercessions that are continually filling the atmosphere of Heaven. That got my attention.

He went on to say that when he heard "Song of the Lamb," he recognized some of the sounds, themes, harmonic structures, intercessions, and spontaneous prophetic singing, which he had heard in Heaven. He felt I had begun to capture something heavenly in the new sound on this CD. He then very simply asked me if I would consider composing a new worship project, which would express the worship of Heaven at a higher and deeper level, taking the lyrics from the Book of Revelation.

At the time that Bob requested this, my schedule was already overcommitted. I had no intention of creating another worship project, and I was already working day and night to complete the Lord's assignments, as well as bearing a heavy travel schedule. Nevertheless, over the next few days, I felt very drawn to this assignment. I love holy assignments, and I am such a songbird, that it really pulled on my heart strings. It went against my better judgment to agree to such a huge commitment for a man in New Zealand that I had never met, and I literally had no time. But I said "yes," not knowing how or when I could start this project.

About a week later, Bob wrote to me and said that the Lord had commissioned him at 3:00 a.m. to write a book about the worship and intercessions in Heaven, and by the next day, he had already written the first chapter. He then asked me if I would write one of the chapters in his book, which would contain spiritual and practical instruction about how to create this type of music, intended to help train worship leaders, songwriters and producers.

Before I had even composed one note, Bob was already enthusiastically writing about the new CD that I "had produced," and how it would bless the global worship community to move into this new heavenly dimension, which will bring the Manifest Glory of the Lord into our meetings.

His exuberant joy and anticipation about this non-existent CD made me feel a bit intimidated for several reasons. First, the Lord has not taken me to heaven, and as that beloved song says, "I can only imagine" what it sounds like up there! Secondly, I had no confidence in my ability to create something as glorious and lofty as Bob was already declaring, as if it were already finished.

The Lord was surely showing Himself strong in this divine appointment. As of this writing, nearing the end of February, 2011, the Lord has already helped me to find the time and the Holy Spirit inspiration to compose all of the songs that will be on this new CD, and over the last four weeks, I have nearly completed the recording process. I can hardly believe that the songs are being birthed before my eyes! But all things are possible to those who love our living, creative, generous Savior.

In this chapter, I will share about how this particular type of song is created and produced on a low budget. But first, we will explore the deepest heart issues, from which these songs must be born. And so, the journey begins!

Special Note to the Reader: Because of my heritage and knowledge of Hebrew, I wanted to let you know that in all of my books and writings, I usually call the Lord Jesus by His original Hebrew name, which is *"Yeshua."* The angel who announced His conception gave this name to Mary and Joseph, and it means "Salvation" in Hebrew.

I know that the Lord will lovingly answer to His Name in any language, to all who call upon Him from a sincere heart. But it pleases the Lord to restore to His Bride from the nations His original identity as the Son of David, the King of Israel. I call Him, "Lord Yeshua" in my private life, and so I will usually call Him, "the Lord Yeshua" in this chapter.

An Old Folkie Grows Up

I've been writing songs since before I found the Lord in 1973. My mother was a classical pianist, and I've been singing for as long as I can remember, as well as growing up playing piano and guitar. After getting saved, I began to write songs about the Lord, obviously. But for many years, the songs were more about me than about Him. I loved to pour out all of my existential confusion and misery into my songs, and I relished expressing them in the most mystical and poetic way possible. The deeper and more mysterious, the better!

As I matured a tiny bit, I started to compose songs based on Scriptural passages. All songwriters who have tried this will admit that there is a fine art of creating biblical lyrics that flow in a poetic way, are faithful to the Scripture, and contain the required melodic, harmonic and rhythmic structures of a high-quality song.

I've written close to a hundred songs, and until recently, most of them followed a fairly standard architecture of songwriting, as it was understood during the 1970's, 80's and into the 90's. As I understood it, there needed to be several verses, which followed one musical pattern, and at least one chorus, which would be repeated after the verses, and which followed a different pattern. In addition, there was sometimes a "bridge," which followed yet another musical pattern, which added some variety before the final, climactic chorus.

I would "massage" the texts in the Bible to get a beautiful flow of poetry, and many times I would combine different Scriptural passages, with related themes, to create a holistic message. In those days, I would rhyme various lines as skillfully as possible. But the rhyming had to be sophisticated, so the songs did not sound trite or amateur. (This was how my mind worked in those days.)

Looking back on my life, I feel that I wrote complex and interesting songs, using biblical themes as a foundation. Technically, I definitely had the skills to write quality songs. But something significant was missing from these songs, which I could not have understood until the time came for the Lord to "awaken love" in me (see SOS 3:5). Even so, I believe that the Word of God is so full of life-transforming power, that any song based on Scripture will benefit the spirit of the one who hears it. Therefore, even in my immaturity and lack of intimacy with the Lord, I was able to write songs that blessed those who heard them, and brought them closer to the Lord.

A Heart Grows Up

After becoming a Jewish believer in Yeshua, my life underwent many changes. I married and in 1982, my husband and I immigrated to Israel, where I gave birth to our three children. After seven years in Israel, we returned to the States, due to some serious problems. In 1994, I received a powerful visitation from the Lord, but it would be ten more years before I would experience His glory again.[13]

I had not yet learned to openly tell the Lord everything I was thinking and feeling, nor had I learned how to wait on Him with an intimate, vulnerable heart, and with faith-filled expectancy. I did not yet understand the bridal relationship with Yeshua that He desired for all of us to experience.

My spiritual acceleration into true worship, dying to self, and intimate bridal love began in 2004, and has increased exponentially since that time. I could no longer write songs that were not birthed from the place of intimacy and from my intense yearning for His tangible Presence.

This heart transformation is the very incubator for the songs that bring an open heaven. Every one of His children can walk in this astonishingly real and passionate relationship with the Lord Yeshua. **And for those of us who have been entrusted with a worship**

[13] The full testimony of my journey in Israel, from depression into intimacy, revelation and glory from the Lord, is detailed in my first book, "Coffee Talks with Messiah: When Intimacy Meets Revelation." Gazelle Press, 2007. www.coffeetalkswithmessiah.com

ministry of any kind, the way Heaven measures success hinges on the purity of the hidden motivations of our heart.

Whether we are singing worship songs to Him, or writing worship songs for Him, the Lord Yeshua is only looking at one thing: Our heart. He is a God of intimate relationship, and He can only receive worship from a heart of sincere, undistracted, passionate adoration.

There are Christian worship songs out there that are huge hits, and have sold millions of copies. They are wonderful, beautiful, glorious songs. And many of them were written by sincere, lovesick worshipers, and the Lord drinks in their offering with deep satisfaction and gratitude.

But the Lord knows the heart of each worship leader, singer, musician and composer. He knows what was in a person's heart when they wrote a song, and He sees what they are thinking about, as they stand on the platform, singing or playing in front of the people. He knows what motivation birthed a song, and He is not impressed with the popularity of a song, nor its appeal to our flesh. The worship that pleases Him is offered "in Spirit and in truth," and this is something only He can discern. Truth is sincerity in the inmost place, without a drop of ego, ambition, pride, competition, jealousy, or desire for fame or fortune.

So, before we move into the actual techniques of writing these uniquely crafted songs, which are in the pattern of the music, intercessions and prophetic declarations of heaven, please take a little time right now and ask the Lord to show you how He sees your heart, when you worship Him in public, or compose songs for Him in private.

A New Song is Born

"Dear Lord, as each precious reader reads this section, I pray that You will accompany these true testimonies with a strong impartation to receive downloads from heaven, which will become heavenly songs. I pray that these words will be quickened by the Holy Spirit to all worship leaders, singers, musicians and songwriters. And for as many as hunger for an increase in the writing of supernatural

songs, would You grant them ever more new music, chords, lyrics and sounds, even as You have granted such treasures of new songs to me? I thank You, Creative Father, for You are pleased to pour out new songs on Your dear children. I ask this in the Name of Yeshua. Amen."

As my intimate encounters, dreams, visions and revelatory experiences increased, I began to write songs that were extremely different from those of earlier seasons. I began to receive songs as I waited on the Lord in the night hours, or as I sat with Him in my "Coffee Talks." Sometimes in my sleep, I would hear a voice singing a line of music to me, and I would wake myself up and write it down, knowing I would never remember it in the morning.

There was one shattering revelation in my life, which was shared in my first book. The process of embracing and contending for this word would be imprinted on my soul. The Lord showed me something very sacrificial and painful in my near future, and I wrestled with it for a long time. Out of the buckets of tears that fell as I struggled to submit to His will, songs would be birthed from that pain and yieldedness. These songs are helping others, who are also wrestling with difficult destinies or Kingdom assignments.

At times in His Presence, I would get revelations, phrases, and free-flowing, non-rhyming poetry, that was like a Psalm of David, but it was from my own heart. I would sometimes get pages of poetry while waiting on the Lord in my rocking chair. By the way, when I say, "I would get poetry," please understand that my mind and creativity were very much engaged, as my spirit would feel these words and impressions. I don't want it to sound like I went into a trance, and behold, a huge poem just popped out. My mind and heart were actively co-creating with the Lord as I would receive these gentle downloads. I was journaling them, no matter how tired I was, or how many chores awaited me.

Not all of these words turned into songs, of course. But with some poetic lines, I would go right down to the piano, and a song would just pour out without much struggle. With other songs, I have had to work very long and hard to craft them into shape. Some songs have been born out of countless hours of trying chords, and rewriting lines dozens of times. Even though it feels less inspired when I work so hard at a song, I know that this process is just as valid, and just as pleasing to the Lord, as the songs that pour out naturally and supernaturally. I believe the Lord is proud of us when we hone our craft to perfection, wanting only the best for Him.

He Beckoned Me!

On March 23, 2007, I received the greatest surprise of my life. (So far!) I was waiting on the Lord in my rocking chair, and He appeared to me. I saw Him with my eyes open, and He looked as solid and real as a "regular person." I had been asking Him a question when this happened, but I can't explain it all here. In this brief encounter, He lifted up His left hand and tipped His head slightly back, and with a quick motion of His fingers, He beckoned me to come over to where He was. I was able to observe the folds of His garments and the expression on His face as He beckoned me.[14]

This gave birth to a very special song and a new CD, appropriately titled, "Beckon Me." During the final days of producing this song in the studio, I had an encounter in my sleep. I heard many angelic voices singing this slow, beautiful descant: "Beckon me to come." It was a contrasting part that would overlay the normal melody that I had recorded. I had not thought of creating that type of harmony yet, one that overlapped the melody, but had nothing to do with the melody. (I didn't know it then, but that angelic download was a little forerunner of the new sound that the Lord would birth in me when I composed "Song of the Lamb," as well as for this new CD, "Sounds of Heaven.")

In the morning, I was having coffee with my daughter Ariela, and I told her about hearing these voices. I said, "They were singing this beautiful part, but I can't remember how it went."

Ariela immediately started singing the part that I had heard! I love that girl. I grabbed her and we raced to the piano, and worked

---

[14] The full account of this blessed encounter is shared in my third book, "The Seduction of Christianity: Overcoming the Lukewarm Spirit of the Church." Destiny Image Publishers, 2010. www.coffeetalkswithmessiah.com

on a three-part harmony that flowed in and out of the melodic lines. Within days, she came to the studio with me, and sang this angelic addition, which made the song more precious to me.

At that time, though, I was still writing songs according to the familiar structure that I described above. I would always build the normal patterns of verses, choruses and bridges.

Remember Me

Still writing songs in this structure, I produced one more CD, "Remember Me." The way these new songs were born is a testimony in itself, because the Lord took me to a new level of revelatory songwriting. I would like to take a few moments to share with you the testimony of "Remember Me," because I am praying that you will feel the Holy Spirit stirring your heart, and imparting new music to you, as I share the Lord's breathtaking initiation of these songs.

The Lord astonished me in early February, 2008. I was six days away from ending my forty-day fast for that year. He commissioned me to write a book on the Feasts of the Lord, which would help His Bride to know and feel His heart of invitation in these feasts. He wanted His people to know they were not just "Jewish holidays," but that He was giving a banquet on His Holy Mountain, and was inviting all of His beloved from the nations to come up and feast with Him (see Isa. 25:6-8).[15]

The Lord helped me tremendously to write this major work in three months. I finished soon after Passover, 2008, and over the summer, I had the strongest desire to write a "theme song" about the Lord's invitation to His feasts, that would complement the new book. I thought of the Lord Yeshua, earnestly desiring to eat that last Passover with His beloved ones. He said, (in my heart's expression,) *"Remember Me, when you break this bread and drink this cup. I am the Passover, and each time you come up to celebrate this feast, I want you to remember My invitation to intimacy. Remember My love*

---

[15] "A Prophetic Calendar: The Feasts of Israel" by Jill Shannon. Destiny Image Publishers, 2009.

and My sacrifice. My last Passover meal was an altar in the presence of My enemies."

He gave me a song so special, that it is as a seal upon my heart. "Remember Me" is a part of the life message and testimony that the Lord appointed for me before the foundation of the world. When I wrote it, I didn't have the slightest intention of producing another CD. I had merely written one special song, and I had no desire to try to work up twelve more songs, just for the sake of having a new project. A few weeks went by.

Then mysteriously, the Lord began to send me His Word. Day after day, week after week, I would hear these sweet requests. Some were from His Spirit directly, and some came as prophetic words from my best friend and intercessor, Cathy. As each request came, the corresponding song would be born within a day or two.

"I'd like you to write a song of intercession for Israel." I wrote "The Embers of Your Name."

"You need to write a song about the power of My blood." I wrote "Manifest the Blood."

"I'd like you to write a song of pure worship to Me." I wrote "Melting Place."

"Would you write the song that Israel will sing to Me, when she recognizes Me? Write it as if you are Israel, who now loves Me." I wrote "Israel's Love Song."

"Write a song about My broken heart for My son." (Here, the Lord meant His son, Israel, but He also was pining for individual sons, whose hearts were far from the Lord.) I wrote, "Where is My Son?"

Then came the most intimate request I had ever heard. *"Wouldn't you like to write a song about the first time?"*

I melted inside, that He would ask me that. "Oh, Lord. You want me to write about the first time You came to me in 1994, when You overshadowed me. Lord, how could I possibly encapsulate that experience into one song?" Although I thought it would be too hard, I immediately wrote "The First Time," which is inexpressibly precious to me. People often write to me about how that song moves them.

Every song was born by a supernatural invitation from the Lord.

All of this happened in a matter of three or four weeks. These songs continued to flow, and I was stunned. That fall, I went into the studio for two or three months and recorded this project. One of the songs was called, "The Day After." It is about the day after the Rapture has occurred, and the Lord had shown me how His heart would feel on that day, and how our hearts (the Bride) would feel on that day.

After I had finished recording this song, my friend Cathy was in her home, listening to the latest CD we had gotten from the studio session. She was listening to The Day After, and when I sang the final line, which was "Waiting for the One who's worth it...waiting for the One who's worth it all," she heard audibly, the Holy Spirit whisper His Name, "Yeshua," in between those two phrases. When Cathy told me what He whispered, I felt the Lord wanted me to go in and record just that one word: His Name, spoken in a whisper, at that moment in the song.

In my next studio session, I went into the booth to record just that one word. I whispered the name "Yeshua," and the song was now complete. I stepped out of the recording booth, and sat down on the couch next to Cathy, as the engineer added this new part to the song. I looked down at my hands, and they were covered with gold dust. It was on my clothes and even on the well-worn couch in the studio. Just whispering His Name was pleasing to Him, and He surprised us with gold dust. I was so happy. I showed the engineer my glittery hands, and he said, "What have you been playing with?"

I was certain the CD was complete. A week went by, and the Lord spoke to me in my sleep. He said, *"Just One Man."* Immediately, I knew that He meant that Moses was just one man, confronting Pharaoh; Elijah was just one man, confronting Ahab, Jezebel and the 850 false prophets of Baal; and the Lord Jesus was just one Man, who would confront sin, death and the antichrist, and He would singlehandedly defeat them. The next day, a friend and I wrote "Just One Man" on the guitar, and it has very unique sound.

Once again, I was absolutely certain that the project was complete. But a few days later, the Lord again spoke to me in my sleep. He quoted me a line from my song, "Remember Me," the song that had started this whole journey. He said, *"Twelve baskets full of crumbs of Your glory."* I knew He was asking for a twelfth song. But I had nothing left to say. I had poured out the treasures of my heart, and I was finished.

On New Year's Eve, I was watching a worship service online with Robert Stearns in Jerusalem. Cathy called me, and I told her that I thought the Lord wanted a twelfth song, but I was empty. She said that she had received a prophetic word a few days before, and had written in her journal the words: "New song?" I asked her to email me the word.

Before the New Year rang in, I had written, "For the Martyrs," using Cathy's prophetic poetry as a foundation for my lyrics. This song rose up from the deepest part of my soul, because it was about the faithful martyrs, the great cloud of witnesses, who were cheering us on into our own destinies. It was intensely personal to my own journey and destiny, and the song flowed out the moment I sat down at the piano.

These songs so deeply stirred and permeated my spirit, that I was certain this would be my last worship CD. I felt I could never exceed the depth of love and revelation that poured out of my heart. But the Lord did not think I was finished.

The Worship Movement

A few months after "Remember Me" was released, the Lord commissioned me to write another book ("The Seduction of Christianity: Overcoming the Lukewarm Spirit of the Church"). This book was to be an urgent wake-up call for the Bride, and an intense spiritual preparation manual for the last days' church. It was about walking in the Fear of the Lord, and cultivating intimacy with Him.

When I completed the writing of this book, I began to think more and more about the nature of biblical worship. I started to feel that just singing worship songs, no matter how inspired, is not enough to offer to the Lord. I began to yearn to create something more conducive to welcoming the thick and weighty glory of the Lord. I desired to write music that would take His people to the very Throne Room of Heaven.

People have so much on their minds, and it takes time to transition from this world into the Holy Place. Worshipers need time to set aside their distracting thoughts, and to enter into the realm of the Spirit. We need to create a worship context that allows us to pour out all that is within our hearts to the Lord. It is hard to create this sanctified context, if we are constrained to a scheduled list of worship songs, immediately followed by church announcements and a sermon.

I have noticed over many years of attending church, that it is very hard to move into deep, prolonged worship in a regular church service, due to distractions and time limitations. I often feel that the Holy Spirit is not given the time or the freedom He desires, to dwell in the midst of His people and to minister to them.

In my desperation to move into the highest praises and the deepest worship, I began to join my heart more fully to the global worship/prayer movement, found in the places of 24/7 worship and intercession. The ministry I am most familiar with is the International House of Prayer (IHOP) in Kansas City, Missouri, under the humble, exemplary leadership of Mike Bickle.

I have been out to visit IHOP a number of times, and I have experienced a depth of worship there that I have never come close to in any local church. As I would spend time in the Prayer Room when I was in Kansas City, or as I watched the live Prayer Room web stream in my home, I observed that their worship teams don't follow the traditional pattern we see in churches.

Their prayer meetings combine worship, intercession and Scripture readings, into a seamless flow of spoken and sung expressions of devotion to the Lord. They use worship songs as a springboard for intercession, the reading and singing of Scriptural passages, and declaring prophetic decrees into the atmosphere.

After all the singers have completed a song, they smoothly transition into singing in the Spirit, to the established chord pattern. Following that, the team moves into a time of intercession, or "worship with the Word." Various singers will sing different lines, prayers or Scriptures; then another singer will respond, and a repetition of a spontaneous chorus might break out. The prayer leader will pray, or a reader will read a line of Scripture, and responsive singing, prayer or declaration will follow. This is called "antiphonal singing." All the while, the musicians are following a consistent pattern of chords that keeps all the singers in the same harmonic structure. At times, they will blend back into the worship song, or enter a new worship song. I too have incorporated this type of spontaneous singing in most of the songs on Sounds of Heaven.

There is freedom in the Holy Spirit for each singer to offer his or her own heart's contribution to the themes that the Spirit is directing. It is responsive, unpredictable, and full of wisdom and revelation.

As we are learning in this book, the reason that this pattern takes us to such depths of glorious worship, is because it is following the pattern of worship found in the Throne Room of Heaven. We will be caught up in the Spirit realm as we worship this way, and the Manifest Glory of God will come into our gatherings.

I have experienced this a few times to some extent, but the one I will never forget happened in the IHOP Prayer Room, on a dark, rainy morning at the 6 a.m. prayer meeting. As the worship team sang, I poured out worship so strongly, that my desperation for the Lord was all-consuming. As I reached up to my God, tears of joy and anguish flooding me, my heart felt ready to burst with my love for Him.

The singers began to sing in the Spirit, sounding like angels, and the moment they did, something happened to me. When my worship reached that threshold, I "disappeared." It was the strangest feeling to go somewhere else, but not know where you went. My body was still sitting there in the Prayer Room, but my spirit went somewhere else, and I couldn't hear the music anymore. After a while, I felt two hands clasping each of my shoulders from behind, although no other person in the room was touching me. These hands remained on me for about twenty minutes. I did not want to move, for fear of scaring Him away.

This expression of heavenly worship will tear down strongholds of darkness, deception and destruction over our cities, regions and nations. This is why the Lord promises to restore the worship of King David to the earth. He was our beloved model of a lovesick worshiper, a courageous warrior, and a prolific psalmist, whose love songs were a never-failing stream of praises, flowing up to the

Throne of God. The Lord is rebuilding David's Tabernacle of Praise, and is now establishing night and day prayer and worship centers around the world.

"After this I will return and rebuild the fallen Tabernacle of David. Its ruins I will rebuild as in ancient, everlasting days, that they might inherit the remnant from Edom, and all the Gentiles from the nations, who are called by My name" (Amos 9:11-12, NKJV).

Special notes to the Reader:

- Before I begin the more technical discussions about writing and recording music, I want to take a moment to address those whose instrument is the guitar, rather than the piano. At this point in my life, most of my songs are composed at the piano, but I have used both instruments to write music over the years. I hope that this section is also relevant and helpful to guitarists, at least to some degree. As I move through these sections, I will try to insert comments here and there, to address how it would be different if guitar is your instrument.
- Some of the instruction I'll be sharing here might be difficult to understand, if the reader is not familiar with music theory. Please don't feel discouraged if I refer to terms or concepts that you have not yet been taught. We are all at different places in our music education, and you don't have to know all the things I'm discussing, to write anointed music. The Lord loves all that you offer Him, no matter what your instrument or your level of music training. You could play four chords on the guitar, and create an incredible song, without knowing all the theory behind it. So if this more technical section is not helpful to you, just skim over it and move on to the next section.

Technical Talk 1: A New Way of Composing Music

My desire to offer the Lord a higher form of worship, led me back to my piano, where I would sit and worship the Lord. Instead of choosing one or another song from the repertoire of worship songs in my head, I began to just play patterns of chords over and over, and started worshiping spontaneously, using parts of songs I had already written, and mingling them with new lines that I had never sung. Guitarists, of course, can also play chord patterns, rather than a "known song," and sing/create spontaneous worship.

One Shabbat (Saturday) morning, I was playing "Israel's Love Song," but I was singing it with changes in the lyrics and melody, as well as singing it in a more free and worshipful way. I went on for quite a while, and I felt the Holy Spirit say, *"While you're at it, why don't you push the 'record' button?"*

I answered, "Lord, it's Shabbat. You know I don't work on Shabbat." ("Shabbat" is the seventh day, which I honor as my date day with the Lord, and He knows I don't normally write my books or record music on the Sabbath.)

I felt like He answered, *"Look at your week. When else are you going to do it?"*

He was right. My six-day work week was crammed from morning till night with the work the Lord had assigned me to do. The demands were so intense, that there was no space for sitting quietly at the piano and recording lengthy parts into the keyboard's memory slots.

I had recently bought myself a long-overdue Yamaha YPG-635 keyboard. I now wish I had done that years ago, but I didn't want to spend the money. It turns out, it has saved me a great deal of money in the recording studio, but I didn't know all that till recently. This beautiful instrument has the capability of letting you record up to five of your own songs (instrumentals, not vocals), with each song having up to five tracks to record, if desired.

The method that I'm describing here was completely new to me. Maybe it is not new to many of you who are reading this, but I had never composed songs this way before. I used to write biblical poetry, according to standard pattern, and then I would create a tune and chords to go with the poetry. This new method involves first recording lengthy musical tracks into the keyboard memory, and later creating Scriptural words, tunes, spoken readings and prophetic prayers, to layer on top of these instrumental foundations.

Guitarists can record audio files of their spontaneous music and singing, using a digital recorder. But they can't record multiple tracks, apart from the audio of their guitar combined with their lead vocal. Of course, capturing the new song in a digital recorder is very helpful, in terms of remembering what they sang and played spontaneously to the Lord, but they would still have to re-record it later in the studio, as well as adding other tracks. They could bring in live musicians to the studio to record the other tracks, or they could bring in a gifted keyboardist, who could synthesize many different instrumental tracks, such as bass guitar, drums, or strings. In my case, the "smart" keyboard is in my home, and it saves me time in the studio, to record as much as I can before starting with the studio.

By the way, if you are a worship leader, but you don't plan on recording your music, you can still use this concept for long periods of live soaking music. The people will springboard off your anointed background chord sequences, whether by singing in the Spirit, offering spontaneous prayers and prophetic declarations, or just soaking in the music and pressing into the Lord. We worship leaders have to set the example of spontaneous praying and singing, because people are shy. We set the tone, by singing in the Spirit, as we play these chord patterns, or we pray spontaneously while playing. Hopefully, others will feel the freedom to join in and express their hearts. If the people stay shy and quiet, we feel a bit "exposed," but I will just keep singing to the Lord in the Spirit, and eventually, a brave worshiper will pick up the baton and run with it.

For those interested in recording into a keyboard, let's go a bit further. These "smart" keyboards come with a software CD, which allows you to transfer the data from the memory of the keyboard into your computer, using a USB cable. It stores the song as a low-memory "midi file," which is easy to transfer to a recording studio application later, like "Pro-Tools."

You can find various chord patterns that sound pretty, and can create a long soaking song, by recording sequences of these patterns, with a consistent rhythmic structure. For example, I might play a pattern of eight chords, and I'll decide the number of measures I want each chord to be played for.

I must enter a time signature and speed before I begin recording; this allows the keyboard to count out and display measures in that time signature, as I record. So, for example, I might choose 4/4 time, and set the metronome (the rhythmic "clicker") to 85 beats per minute. If you record without using the clicker, it will be much harder when you go to import it into the studio's application. The recording engineer will have a hard time cleaning up your tracks, cutting and pasting, or cloning good measures. He really needs your measures to be even and consistent. You might feel constrained by the relentless beat of the clicker, and you wish you could slow down or speed up at various moments in the song, but in my opinion, it is worth the discipline of the clicker, because of how easy it becomes later to adjust many things in the studio.

Then I would play, for example, this sequence: Bm G D Em Bm Em D F#m, playing each chord for only one measure. Perhaps I would repeat the first four chords four times, and then I'd repeat the second four chords four times. If you only play each chord for one measure, this would give you 32 measures. Then I might repeat this pattern twice. Or I could create a second and a third new pattern of chords, and I could record them in sequence after the first pattern. **Of course, you must make sure your patterns are all in the same key**, unless you skillfully and deliberately modulate the song. Otherwise, it will be musically chaotic, and will not hold together as a true song.

These sequences could end up being an instrumental section within the song, or they could be the background track for lead vocals, which I would later compose and record, using verses of Scripture, or words from my own heart of love. If you leave them as instrumental background, you can later record spoken Scriptures over this music, or you can pray prophetically over this music. It is a wonderful tool to impart to the listener a vast download of Scriptural and prophetic truth, using the music as a foundation.

Before you record your first track, which would often be a "piano" voice, you should write down your chord sequences, and mark how many repetitions of each pattern you want to do. Keep recording until you feel the song is the right length. Then, begin to slow down your playing, and turn off the clicker, while wrapping up the last few measures at a slower tempo. Or, just ignore the clicker, as you finish.

Once you have finished the first track, you can overlay other tracks. I often record the mellow "synth pad" track after the piano track, as a smooth filler, followed by the bass guitar track (these are all synthetic instruments, but they sound absolutely real, at this point in technology's development of synthesizing real instrument voices). After that, I might record a creative orchestral part, cello, flute, air horn or human choir track. These choices depend on the feeling and flavor of the song. Each time you add a new track, you are listening to the playback of the tracks already recorded, so it is quite easy. The keyboard displays each measure during playback, so you can identify which measures contain mistakes, and later, when you go into the studio, you can easily fix these. You can't fix mistakes in the keyboard, without having to redo the whole song. I always fix mine when I get to the studio.

One more important point: the "voices" that your keyboard produces are not the same as the choices of voices that your engineer in the studio will have. So even if your "cello" at home sounds fake, when you take your midi file to the studio, the engineer can choose from many true cello voices, that are actually real, recorded, instruments. It is amazing what synthesized instrumentals have become.

When the Holy Spirit said, *"While you're at it, why don't you push the 'record' button?"* this was the process I went through on that Shabbat morning, and this is how my last CD, "Song of the Lamb" was born. By the end of that day, I had two long soaking instrumentals, each with very different sounds and qualities. One would become a prophetic version of "Israel's Love Song," and the other sounded like a lush, romantic movie score, just waiting for some words to go with it.

When I played the romantic "movie score" instrumental for Cathy, she said it sounded like the Song of Solomon. I did not know how to write lyrics from the Song of Solomon, because I had never really understood that book. At first, I tried to just mechanically pull out verses and create lines that went with my music, but I got frustrated very quickly. I was working on these lyrics as if they were a difficult chore, and I was missing the Lord's heart.

A few days later, I was having my "Coffee Talk" with the Lord, which is a time of intense intimacy and talking with the Lord. In my heart I was thinking, "When I finish my coffee talk with the Lord, I'll go downstairs and start working on these Song of Solomon lyrics again."

The Lord heard my thoughts and said, *"Why are you trying to 'work' on this song apart from Me? Why are you treating it like a chore that you have to work on? The only way these songs can be birthed is in the place of intimacy."*

I wondered what the Lord was saying, in this gentle challenge.

The Lord said, *"You can go off later and work on them alone, or I can just give them to you right now."*

This was too good to be true! I looked up and said, "You mean I can just get out my Bible right now, and You will tell me how to place all these verses in the song?"

In faith, I got out my Bible and notebook, and the Lord helped me to find verses, and He guided me, as to where to place each line. They matched with the musical sequences so beautifully. Later, I added a spoken dialogue between the Bride and the Bridegroom, and added spontaneous prophetic song and spoken prayer. This became "My Garden," the utterly romantic song of all songs.

Technical Talk 2: A New Way of Composing Harmonies

In composing the "Song of the Lamb" project, the Lord showed me which Scriptural themes He wanted for each of the five long songs on this project. I began to write lyrics to all the songs, carefully working through the poetic aspect of every line.

I went into the studio, and began the process of recording all the lead vocals, harmonies, Scriptural readings, spontaneous singing and spoken prophetic prayers. I finished recording the complex first song, "Song of the Lamb." At least, I thought I was finished.

That night in my sleep, I heard a man's voice singing to me: *"Oh, You were born, and You died, and You rose again to life."* Instinctively, I knew the chords that would go with this line, and I got up to write down these words, along with the tune and chords. Then I went back to sleep. A few hours later, I heard the most beautiful female voice I have ever heard, singing, *"You told me You loved me a thousand times. Your banner over me is love."* I got up again and wrote it down, words, tune and chords. I knew that these words would follow a similar chord sequence to the earlier line I had heard. This chord sequence did not yet exist in the "finished" version of Song of the Lamb.

The next night, I heard a voice singing, *"Let us taste and see that the Lord is good. He has rescued us from our brokenness."* It would also go with the new chord sequence. At this point, I realized that the Lord was giving me a new way to write harmonies: Singing a collection of different musical lines, which would intersect beautifully, when sung to a particular pattern of chords. I wouldn't know how all these lines would sound together, until I went into the studio and tried it!

I continued receiving musical downloads for this song, and I knew that the Lord was creatively giving me a different ending sequence, which would replace the one I had already finished recording. So I went back to the piano and recorded the new chord patterns that the Lord had shown me. I took this new "ending" instrumental file on my flash drive into the studio, and we deleted the last five minutes of the song, as I had previously recorded it. I asked my engineer to import this new sequence, and paste it in as the new ending.

I then sang the new lead vocals, which consisted of the five or six lines that I had heard from the Lord. I could only sing one line at a time, of course. (Are you disappointed in me that I can't sing all those parts at the same time?) Then the engineer looked at me and said, "So what am I supposed to do with all these lines?"

I answered, "Paste them all overlapping each other, on top of the new instrumental, and let's see how it sounds." I also asked him to paste in the original melody, "There is One seated on the throne, like a Lamb that was slain," over the new chord structure.

He looked at me like I was crazy. Then he shrugged and pasted my many lines, one on top of another, over the new chord pattern we had imported. There wasn't one bad note! They sounded like multiplex harmonies, with many unrelated lines flowing in and out, without conflict.

I was truly awed at how the Lord helped me to enter a new dimension of songwriting. I felt childlike joy and wonder each time I would hear all of these parts blending together. Not only did they sound beautiful, but the Lord Himself was my musical coach! That thought brought me great happiness.

When I was completing the last song, which was an expanded version of my earlier song, "A Part of Me," I created a long space at the end for this new type of mutliphonic harmony. I went to bed each night, expecting the Lord to sing to me these new lines in my sleep, as He had done with Song of the Lamb. After a few nights had gone by, I hadn't received even one line in my sleep. I was disappointed. As I was waiting on the Lord, I felt His Spirit say to my heart: *"I showed you how to write this type of music. Now you try it."*

Oh, of course! He is my coach, and He has shown me an example of how to compose these lines. Now, like any good music teacher, the Lord Yeshua wanted to watch me create this type of harmony for the final song. I felt honored to be His student, and I began to create lines that would intersect in a beautiful way, flowing with the chord sequence of the ending.

I managed to compose about four lines that worked in the ending sequence, and couldn't think of anymore. I went into the studio and recorded them, and they sounded wonderful. Over the next few days, the Lord sang to me about three more lines that just added to the harmonic structure of the climactic ending of this most intimate song. One night I heard someone singing, *"Self-absorption is a sin. Look to Him and not within."* I was surprised, because this line wasn't exactly the theme of the song, but I wanted to be obedient, and so I sang it as part of the ending sequence. He was faithful to put the icing on the cake, but He wanted to wait until I had given it my best effort.

While recording "Kadosh," I heard this line sung in my sleep: *"Under the stars, under Your arms, under the heavens, I wait for You."* Again, I found the perfect spot to sing this overlapping melody with the original chorus.

Technical Talk 3: Prophetic Singing

It is important to leave instrumental spaces in these songs for spontaneous singing, whether in our own language, or in the language of the Spirit (also known as "tongues"). It is easy for us to understand spontaneous singing, when it is in the context of live worship. Live worship is often a spontaneous experience, at least to some extent. People are free to sing lines that the Spirit just puts in their heart at a given moment. But what about in the recording studio? This is a huge leap of faith!

Under the pressure and expense of recording in the studio, how do you just go off in the Spirit? There is an engineer sitting there, and you are not in a glorious worship environment. If you have other singers, can you trust them to just start singing lines that will blend with the others? And which parts will be worthy to end up on a "polished" project? And if each singer does three different spontaneous "takes," how many paid studio hours would it take to sort through all of their offerings, and blend together the perfect "spontaneous" combination?

All of these concerns were in my mind when I recorded "Kadosh," the second song on Song of the Lamb. I had left a 32-measure section of instrumentals in the center of this fourteen-minute song, and I wanted my male vocalist, Michael Hendricks, to just sing whatever was in his heart throughout this section. I also wanted my daughter Keren to sing into this section, although they would not be in the studio at the same time. I prayed for Michael as he went into the room to sing his part. I told him, "You can sing in English, or in tongues. It doesn't have to be the exact theme of this song...just sing anything about the Lord that comes to you."

He went in and just sang the most beautiful, unrehearsed part. The melody he sang was completely different from the regular Kadosh melody, but it fit the chord pattern perfectly. It was flawless, and we didn't need to do anything to adjust it or improve it. I am so proud of this awesome young worship leader!

A few weeks later, my daughter Keren came into the studio, and recorded her spontaneous, angelic part. The way her high, airy voice blended with Michael's rich baritone was marvelous. Many people tell me that this is their favorite section of the song.

I also sang some unrehearsed prophetic parts on this CD. I had never done this before, and I was nervous. But the Lord just filled me with lovely, creative parts that had never entered my mind until the moment I stood at the mike. When the Lord is pleased with the heart of a worshiper, He will fill our mouth with good things!

There is something very special about stepping out in faith and singing an unrehearsed, unplanned part into the mike, whether live or in the studio. I believe the Lord is pleased with us when we try something new, which requires trusting Him implicitly to put something precious in our heart and in our mouth, something we never worked on ahead of time. It's like letting go of our control and walking on water. Those who trust in Him will never be put to shame.

If you or the other singers are willing to be courageous, the Lord can bring out some wonderful surprises in your live worship or your recorded projects. We all mess up, so I'm not saying all our surprises will be happy surprises! That's where the humility part comes in. It might not always be perfect the first time we try, but something happens in your spirit when you sing in the Spirit, or sing spontaneously in your own language. It enlarges your inward capacity to trust the Lord and to receive revelation from Him. I really believe that, because I tried it, and He was beyond faithful.

When I recorded this new CD, "Sounds of Heaven," my engineer put three of my singers in the same large recording room, each singer separated by a glass door, and all hearing each other on headphones. In most of the songs, I had left sixteen or thirty-two measures of instrumentals, near the end of the song, specifically for singing in the Spirit. And for each song, the singers would start singing at the same time, each one just singing a long, spontaneous part, over my chord structures. They all sang something totally different, and yet it sounded awesome together. The results were far beyond what I "accomplished" on Song of the Lamb, where I was mostly singing parts with many other "Jill's." I'll cover more about this in the Sounds of Heaven section.

Technical Talk 4: Spoken Parts

If you have left some open spaces of instrumentals without singing, you can have readers, speakers or prayer leaders speak over these sections of music. This applies to live worship, as well as recording songs. **It is very powerful for the listener to hear Scriptures read aloud over music, or to pray heartfelt prayers or prophetic declarations over this anointed music.**

When you read a passage of Scripture, try to adjust the timing of your reading to fit nicely into the timeframe of the music. (This is only in the case of studio recordings. If you are in live worship, you can play as long as the prayer leader needs.) Don't rush, but read the words with feeling and pause when needed. You can combine different verses to form a theme you wish to read. Sometimes I will choose two or three passages, and I will only choose certain verses from each one. Then I'll read them as a seamless piece, to create the theme I wish to emphasize. You can be creative with the Scriptures, and you don't have to read a long passage, if all you need is a few verses to make your point.

If you desire to pray, make sure it is a prayer from the heart, even if you compose it ahead of time and write it down. That is no problem, as long as it is really a prayer that you would pray privately before the Lord. Don't use overly religious language if you pray, but talk to the Lord as you would talk to a beloved and revered friend. Make it real, and it will be powerful in your worship service or CD.

I need to write down my prayers ahead of time, when I'm planning to record them during a long song. I practice them at home to the track, so I can time my words perfectly, and maybe adjust the number of words, if the timing isn't working out. Then I read it in a fresh way that sounds spontaneous, because it is still coming from my heart.

Sounds of Heaven

As you may recall, Bob Misst requested that I produce "Sounds of Heaven," based on the many types of heavenly sounds that he heard in the worship in heaven in his vision. That's why I took the time to explain my process in creating "Song of the Lamb," because this new project has employed the same musical strategies and architectural principles.

As of this writing, I have written all the songs that will be on "Sounds of Heaven," and I have almost completed recording my instrumentals and vocals in the studio. I have spent about twelve long days in the studio so far, and the songs are close to completion. If I had not been able to record most of my instrumental tracks at home in my Yamaha, this accelerated progress would have been impossible. For the first six sessions, we cleaned up my instrumental tracks, added percussion, and put down all lead vocals and most of the multiple harmonic lines. I sang many harmonies, according to the patterns I have explained in this chapter, as well as many "normal" three-part harmony lines.

Before my other singers came in, the music sounded like there were about seven of me! The problem is that heavenly sounds require a diversity of voices. It requires male and female, Jew and Gentile, diversity and unity. I desperately needed other singers, other types of voices, and I needed some good male voices in particular. The Lord knew my heart's cry, and He provided me two precious girls (my own daughters) and two special young men of God, to enrich the sounds of this project.

Three of my singers were my daughters and their awesome friend, Charles, a gifted musician and singer. These three singers were very comfortable improvising together. When they all stood in their recording spots, each one singing a long, unrehearsed part to these long chord sequences, I was absolutely stunned at what I heard. The way their voices blended, responding to each other, and weaving in and out like a "tapestry," I was amazed. I prayed in tongues in the control room, for the whole time they were recording. They were all singing about various themes found in each song, and it came out just so fresh and wonderful. I usually placed the singing in the Spirit at the end of the songs, as a way to take the listener to the deepest place, as the song was ending.

I knew the Lord was giving me this gift, to make this CD very special. It adds a dimension, whereby the listener has some time and space to go off into worship, and to add his or her own melodies with the free flowing singing in the Spirit. It steps off the "map," and just opens up each heart to sing whatever flows out at that moment. I cannot tell you how grateful I am that the Lord gave me singers who could step into this dimension, as they are all very familiar with the worship at IHOP. Michael also came in later, and graciously added his own beautiful parts, during a separate recording session.

When I began this project, I wrote out the passages in Revelation, from which I would be taking most of the lyrics. Bob has presented these Scriptures in this book, so you know most of the sections I used. I also found other Scriptural passages that helped me express these biblical themes, so you will find these throughout the songs. Some are from Psalms, or from the Gospels, one of the prophets, or other sections of Revelation. And of course, there were places where I wrote my own words, from my heart of love. He wouldn't have it any other way, beloved songwriter!

I sat at the piano and began to create one song at a time, trying many chord sequences, as I described earlier. Then I tried to see which passages from Revelation would best match each chord structure. Most of the songs in this project each contain three or four different chord patterns, which are mixed and matched in a variety of ways. In the same way that the Lord taught me how to write intersecting harmonies in Song of the Lamb, He also gave me the same types of harmonies in "Sounds of Heaven."

What the Lord taught me in Song of the Lamb matches perfectly with what Bob has written, concerning what he has heard in heaven. He heard many voices singing different choruses, each line combining with, complementing, and intersecting each other. One line is sung by the Cherubim, another is sung by the Elders, yet another is sung by the creatures on the earth, and another is sung by the redeemed in Heaven.

Do you see how perfectly the Lord had prepared me to follow the pattern that Bob saw, before I ever knew Bob and before Bob ever knew me? The Lord trained me ahead of time, teaching me how to write intersecting lines of music and lyrics, which create a "perfect" symphony of worship on earth, as it is in Heaven. When I say "perfect symphony," please understand my heart. We can only pattern heaven's music in our own very limited, weak, finite way. I am all too aware of my physical and technical limitations, never mind my budget! But, as Bob wrote, the Lord doesn't need perfect music on earth. He needs loving, yielded vessels, who will worship Him with all their heart, and will offer Him their very best. **Our best is ALWAYS good enough for this most affectionate, loving, gentle, patient Friend and Lord and Savior and Teacher and King that we are privileged to serve.**

When the Lord began to teach me, He sang new lines to me Himself, straight from Heaven. And so, when I received that prophetic, destiny-laden email from Bob, it was like I had already been prepared for this new assignment, with every building block already in place. Had the Lord not gone ahead of me on this, I never would have been able to do what brother Bob needed for this duet-project.

How awesome is the Holy One, who was and is, and has gone before us, behind us, underneath us (like a mother eagle), and into our future. Oh, how He has us covered, beloved ones of the Bridegroom! How He has hedged us in, behind and before. Such knowledge is overwhelming to me, and my mind cannot wrap around this eternal truth!

Technical Talk 5: Sound Effects

One of the most intimidating things that Bob asked of me was to create some of the "sound effects" he heard in heaven. I was very afraid to even touch this area, because I am not gifted in this field, and I knew that it would be impossible to duplicate what he heard in heaven. I didn't want it to sound like a theatrical "show." I didn't want to clutter up the music with too many sounds. I also knew that it could add much time and expense to the production of this CD, which frankly, I could not afford. So this part really frightened me, because I didn't want Bob to be disappointed with what I was able to produce. This dear man had such high anticipation and expectations of this project, and I felt like one small person, without a band or singers (at that time). But the Lord can do SO much with so weak a vessel, who is working only for His glory.

As of this writing, I have added a small number of "sound effects." They are measured, and are not used too often, but I think they add a nice touch. "Less is more," when it comes to sound effects in worship. Additionally, my singers and I have recorded some moments of spoken whispering, speaking and shouting the praises of God, which we are inserting in strategic places under the music. This will not sound like millions of angels, of course. It will sound like three or four normal people, praising God. But the Lord will smile when He hears little us!

As an added blessing, one of the singers who helped me is a gifted musician and multi-media technician from the International House of Prayer in Kansas City, Missouri. His name is Charles Vadnais, and after he arrived in Philadelphia to sing on this project, I learned that he is an expert in sound effects! We only had a few hours to work on this aspect, before he had to leave town again, but he was able to teach me a lot in that time. He created four or five "effects" for me on my computer in about four hours. We used a program called Adobe Premiere Elements 9, which I had just purchased and installed last fall, but didn't know how to use. I had wanted to use it to edit my teaching videos, so that I could put edited video teachings up on my website, but it didn't come with any User Manual, and I was not able to figure it out by myself.

Charles saved me so many studio hours by creating these strategic sounds on my own computer. Bless him, Lord. They are fitting in wonderfully with the songs. They are subtle, yet helpful. I want to be sensitive to the listener, who might be deeply lost in worship, as they listen to this peaceful, rolling music. I don't want them to be suddenly jolted by the loud clap of a thunderstorm, or the shrill blast of a trumpet. I don't want these sounds to distract or detract from the beauty of what the Lord has given me. Sound effects must only be used to enhance the worship experience, not to detract from it with showmanship.

As Bob wrote so beautifully, our main emphasis must be worship. The Lord knows that we cannot possibly create the infinite sounds of heaven accurately, and He also knows that what is freely produced in heaven is expensive to produce on earth. Therefore, I will only have a minimal amount of thunders, rushing waters, voices and babbling brooks, adorning the music on Sounds of Heaven. And I now know that neither the Lord Yeshua, nor our dear brother Bob, will be disappointed in my small sounds, because the heart of worship has been fully expressed.

A Practical Example

To close this chapter, I'd like to give a practical example. Just as Bob included the lyrics to my "Song of the Cherubim" in this book, I've chosen another song from Sounds of Heaven, which will be the fifth track on this CD, to share with you.

I'm presenting here the lyrics and chords, and when you listen to this song, you will better understand how I constructed these overlapping lines. When I typed out these intersecting lines, I deliberately spaced the words out, so that you could see how these two lines are synchronized. I wanted you to see which chord lines up with each word in parallel lines. You will also hear spontaneous singing by my other singers, and these lyrics are not printed here. It would not be possible to write out all the words which my singers spontaneously sang into these songs, so I can only include the "official" lyrics.

Salvation Belongs to our God/The Marriage of the Lamb
Words and Music by Jill Shannon © 2011

Read Rev. 7:9-10

```
D          A(C#)          Bm
Salvation belongs to our God
        G                D A(C#)        Bm   G
Who sits upon the throne and to the Lamb (4X)
```

Read Rev. 7:13-17

```
D                Em           D(F#)           G
All blessing and glory and   wisdom,   thanksgiving,
       D        A(C#)         Bm   G
And honor and power and strength
```

```
Salvation     belongs to our God,      who sits upon the
Throne          and to the    Lamb
```

New Chord patterns to move into Wedding song:
Read Rev. 19:6-10
Bm G D Em (4X)
Bm Em D F#m (4X)

```
Bm                G                   D                    Em
Halleluiah, halleluiah, for the Lord omnipotent reigns (4X)

                  Bm                Em
So let us be glad and let us rejoice
                              D                         F#m
For the marriage of the Lamb                            has      come
                  Bm                Em
For the wedding of the Lamb has come
                              D                         F#m
And His                       Bride has made herself ready
```

Read Rev. 19:11-16
Bm A(C#) D Em D(F#) G
Bm A(C#) D Em D(F#) A

```
     Bm      A(C#)   D
His eyes are like a flame of fire
        Em       D(F#)     G
On His head are many crowns
        Bm       A(C#)      D
He has tread the nations in His wrath
        Em          D(F#)   A
He has struck the nations down
     Bm        A(C#)       D
Gird on Your sword, O mighty One
       Em       D(F#)  G
Ride forth victoriously

        Bm       A(C#)      D
You are King of Kings and Lord of Lords
```

```
        Em        D(F#)             A
Let the earth be filled with Your glory


           Bm                    Em
So let us be glad and let us rejoice
                          D          F#m
For the marriage of the Lamb has come (4X)

   Bm             Em
Halleluiah        Halleluiah
                  D                     F#m
For the       Lord omnipotent        reigns (4X)
Bm                              Em
For the wedding of     the Lamb has come
                  D                           F#m
and His      Bride     has made herself    ready
```

Continue same choruses as above, plus this below, but with these chords

```
Bm              G               D                 Em
Halleluiah, halleluiah, for the Lord omnipotent reigns (4X)
```

A Blessing from the Lord and from His handmaiden

I am honored that Bob gave me the privilege of writing this chapter for you. I know that each of us write songs in our own unique and marvelous gifting from the Father of Lights. No two worshipers have the same heart or the same language of musical expression. We are as unique as snowflakes, and our diversity pleases the Lord greatly. He is waiting to hear your voice, beloved. He moves to the sound of your voice. **Take time aside to minister directly to the Lord, in your own private and secret time and place of worship. This will conceive intimacy, which will give birth to the songs of heaven for the Bride.**

I was asked to share my particular journey and process, as a way of instructing, stimulating and inspiring you to offer your own language of love to your Beloved. Do not be concerned if my technical process, equipment, or budget is not available to you. There is always a way to capture the beauty of the sounds of heaven, when you are creating music only for Him.

If my techniques of writing lyrics or harmonies are not helpful to you, that is fine. You will be developing your own new heavenly strategies for writing songs, and I hope that I will too. I have a classical/folk background, and my songwriting is heavily influenced by these genres of music. You may have a rock or blues background, or even no background. Perhaps you have just learned your instrument, and are writing very simple, beautiful love songs to our dearest Yeshua. Perhaps you've never learned music theory, you don't yet know how to write 3-part harmonies, and what I wrote is "Greek" to you.

Please do not worry about it. I can only be who I am, and I'm expressing my musical journey, hoping it will bless you and help you grow as a musician. But I would be heartsick if I knew that my level of instruction has caused anyone to feel inadequate as a musician.

The most important thing is that we have access to the sounds of heaven, through the blood of Yeshua, and through our intimate worship and adoration to the Lord. As we lose ourselves in lovesick abandon, the Spirit of the Lord will bring to you the sounds of heaven that He desires to hear from you, only from you. No one else can fill the place in His heart that YOU can fill.

Know how much you move your Bridegroom's heart with one glance of your eye, or one melody from your lips. Your offering will move Him to tears. He is comforted by your tender and childlike expressions of love. He drinks deeply of your love like the finest wine, and He will remember your love more than wine. He will remember your love when you stand before His throne, beloved.

Never stop singing and making music to the Lord! They don't stop in Heaven, and we must worship on earth, as it is in Heaven.

To the praise of His glory and grace, I bless you in the Name of the Father of Lights, who sits upon His throne of love. And in the Name of the Bridegroom of the ages, even Yeshua, who has saved Himself for His wedding day with you. And in the Name of the sweetest Breath that has ever breathed songs into our lungs, the *Ruach HaKodesh*, the Holy Spirit of God Most High.

Your loving sister,

Jill

PART II

The Pattern of Intercession As it is in Heaven

CHAPTER 5

PREPARING THE EARTH FOR THE RETURN OF THE LORD

In Part I, I have explained worship as it is in heaven, as I experienced it in the vision of the throne room. I mentioned earlier that it was not a new revelation, but was akin to the worship described in the Book of Revelation. In the first chapter, I shared various prophecies that were given at different times, as well as the vision of worship, which the Lord gave me. The Book of Revelation gives us a glimpse of the worship, intercession and warfare, which are going on together at the throne of God. In Part II, we will focus on intercession in the throne room.

Worship, intercession and warfare at the throne of mercy, has an ultimate purpose. **The Father's purpose for creation was to prepare a bride, which is the church, and a dwelling place, which is the earth, where the marriage of His eternal Son and the bride will take place.** The worship, intercession and warfare have this one specific purpose in heaven: the restoration of creation and renewal of the earth.

If the church in heaven has this one specific purpose, then the core of the church's activity in these last days should be joining with heaven in this process of restoring and renewing the earth. Heaven's purposes should be the church's purposes. The concerns of the Father's heart should be the concerns of his children's heart, the concern of His true church. What we see in the Book of Revelation is the key: intercession and warfare in worship, as heaven's agenda to restore and to renew the earth. This should be the same key for the church.

Why Focus on Genesis and Revelation?

At the heart of Christian hope, is the assurance that after the death of our earthly bodies, our true life continues with the Lord in heaven. The bible indicates that the state of this "blessed hope" (which is heaven), is only for a period of "time." Although heaven is our hope, the Spirit and the bride cry out for the Lord's return to earth. "Christ has died, Christ has risen, Christ will come again," is a well-known declaration of the Christian faith used repeatedly in many church services. All those believers presently with the Lord in heaven, according to 2Cor.5:1-8, will have their spirits/souls reunified with their bodies and will dwell on earth. God will dwell in our midst as He had intended it to be in the Garden of Eden. This is the Joy of God. This is His desire.

In the Apostles' Creed, most churches are familiar with the line, "I believe in the resurrection of the body…" Christians not only believe in the immortality of the human soul, but also that the human body is destined to rise to immortality from the grave. The final hope of the Christian is not to "go home" to the Lord. It is for the Lord to "come home" to us. The purpose for the Lord giving us a new, resurrected body is not because we need one in heaven. It is on the earth that we need our glorified bodies, in order to reign forever with Him. Sin has corrupted this body to experience death and decay. Our hope is for Him to return to the earth during our lifetime. For those of us who will die before His kingdom is fully established, which occurs at the second coming, we can look forward to our presence with God in His spiritual abode of heaven. Yet even in this blessed state, we will not be fully satisfied, because God has more in store for us. Those believers who are presently with the Lord, will have their spirits/souls reunified with their bodies, and will dwell in the new earth with God. He will be in their midst, as He intended it to be in the beginning (1Cor.15:22, 53; 1Thess.4:16-17, see also Job

19:26-27). Let us continue to look in hope for the coming kingdom prepared for us by the Lord Jesus Christ Himself!

The first book of the bible, the book of Genesis gives us a glimpse of the pristine earth that was lost. The last book of the bible, the book of Revelation tells how this once pristine earth "that was lost" will be restored. The restoration and renewal of the earth is essential to the return of the Lord Jesus Christ. Warren Lyon's prophetic word, "I will walk with you in the snow as I walked with Adam in the garden," has restoration and the renewal of the earth as its core message. With the second coming of the Lord Jesus to the earth, He will establish His kingdom reign of righteousness, justice, peace and joy. This is a stark contrast from what it is today. Not one government, or one president, or one monarch, has achieved this perfection during their reign, nor will they ever. After the voting activity decides the leaders, power gets to their heads. They turn from servant-leaders to slave drivers! How different from the leadership of our Lord Jesus, who laid down His life for us. Surely, His government only can bring the joy and peace our hearts yearn for.

If the church longs for the return of its Lord and Master, then this longing should draw the church into worship, warfare and intercession, which will hasten His return. This heart's cry should not merely occur once a year, as a Christmas meditation, or a Good Friday reflection. The bride's longing must be expressed continually, as she cries out for the return of her husband, her Lord, to the earth. The church needs to join heaven in this restoration and renewal process through worship, warfare and intercession. The Book of Revelation, along with the apocalyptic prophecies of the Lord Jesus found in Matthew 24, Mark 13, and Luke 13; speak about the process of restoration and renewal of the earth.

The book of Genesis is the book of many beginnings. In Genesis, we find the beginnings of the creation and the earth, the dwelling place of man. In the opening chapters, God and man walk together in harmony and balance, and in an intimate relationship, which reflects the image and likeness of God his maker. This relationship distorted and twisted by sin. In handing over the "title deed" of the earth to the evil one, man's relationship with God is effectively perverted. This distortion then corrupts Adam's relationship with Eve, his helpmate, as well as corrupting the entire created order of Genesis 1.

God's plan for the restoration of man and earth finally come to their consummation in the last book of God's Word, the book of Revelation. Genesis and Revelation comprise the foundation of how we understand why certain "things which must soon take place" (Rev.1:19). Why does this earth, so tainted by sin, needs restoration and renewal prior to the Lord Jesus' return to the earth? What must the earth look like, when the Lord returns for His bride the church, at the marriage feast of the Lamb, as revealed in the book of Revelation?

In Genesis 6:5 to 8:22, we encounter the Midrashic pattern again. We see that the judgment of God destroys every living thing outside of Noah's Ark, because of the extent of sin. When Noah and his family came out of the ark – it was a "new world." The flood had destroyed the old order of life. **In the book of Revelation, we see that God releases judgment on the earth in response to the worship in heaven, in union with the worship of the bride of Christ, the church on the earth.** The effect of God's judgment brings about a "new earth that comes down from heaven," a restored and renewed earth, born out of the destruction of sin. Revelation 21:5 says, "*And He who sits on the throne said, 'Behold, I am making all things new.'*" *And He said, "Write, for these words are faithful and true.*" The Greek word translated "new" in Revelation 21:5 can also mean "renewed" or "restored." It is the same word used in Revelation 21:1, referring to a "new heaven and a new earth." Again, in Revelation 21:1 which is a reference to Isaiah 65:17, the new heavens and a new earth are mentioned. The Hebrew word for "new" in Isaiah 65:17 is derived from a root meaning to "rebuild, renew, or repair". The Lord Jesus described His second coming as the renewal of all things:

"*Then Peter said to Him, 'Behold, we have left everything and followed You; what then will there be for us?' And Jesus said to them, 'Truly I say to you, that you who have followed Me,* ***in the regeneration*** *when the Son of Man will sit on His glorious throne, you also* *shall sit upon twelve thrones, judging the twelve tribes of Israel'"* (Matt. 19:27-28).

The Lord Jesus, in answering Peter's query, brings into focus His return to set up His earthy kingdom, over which He will reign. His kingdom would come despite the rejection from Israel's leadership. His kingdom will fulfill the Messianic restoration that Israel hoped for. Thus, the earth will experience:

- Spiritual restoration (see the Messianic prophecies of Isaiah 2:3; 4:2-4 and 11:9)
- Political restoration (see Isaiah 2:4; 11:1-5, 10 and 11; 32:16-18)
- Geographical and physical restoration (see Isaiah 2:2; 4:5-6; 35:1-2)[16]

When all these restorations are complete, the Lord Jesus will then sit on His glorious throne, and the apostles will receive their reward.

Peter, under the power of the Spirit of God, in his first Pentecost sermon, speaks about the same Messianic restoration (about which the Lord Jesus spoke), and uses the same word "restoration."

"Therefore repent and return, so that your sins may be wiped away, in order that times of refreshing may come from the presence of the Lord; and that He may send Jesus, the Christ appointed for you, whom heaven must receive until ***the period of restoration of all things*** *about which God spoke by the mouth of His holy prophets from ancient time"* (Act 3:19-21).

It is an interesting comparison when one takes the record of what was lost in Genesis 3, and what is restored in Revelation 21 and 22. Here are a few:

---

[16] Barbieri, Louis A., Jr. (1983), "Matthew," *The Bible Knowledge Commentary – New Testament,* Wavoord & Zuck, eds. (Wheaton, IL: Victor Books), for a deeper exegesis.

"Paradise Lost" in Genesis	Restoration in Revelation
After Adam and Eve made a choice to follow satan's deception, God pronounced a series of curses that would come upon His creation, as a result of the Fall.	The last few chapters of Revelation reverse the curse of sin upon man and creation, which started in Genesis 3.
1. The ground was cursed (Gen. 3:17); it would become the cause of disagreement between nations, between peoples, between families.	1. The curse on mankind and creation is removed (Rev. 22:3)
2. Sorrow would come upon the human race (Genesis 3:17).	2. There is no more sorrow and pain (Rev. 21:4).
3. There would be a struggle to survive and grow food, in the midst of thistles and thorns and hard toil (Gen. 3:18-19).	3. Every tear from man's eyes is wiped away, meaning that hardship and struggle to live will be taken away. The earth and all creation will produce abundantly, as it was designed to do. (Rev. 21:4).
4. Death and decay were man and creation's reward – man would return to the ground, from which he was created. (Gen. 3:19).	4. Death is banished in the renewed and restored world in Revelation (Rev. 21:4).
5. Adam and Eve were given coats of skin to clothe themselves (Gen. 3:21).	5. Animal coats are replaced by clothes of fine linen. (Rev. 19:14).
6. Satan and his fallen angels, called demons, would harass humankind and work towards their destruction and that of creation (Gen. 3:15).	6. Satan and the evil minions have been destroyed prior to the recreation of the earth, and are not present in the renewed earth (Rev. 20:10).
	7. Knowledge is not used for the exploitation of the world's resources, in order

7. Adam and Eve were cast out of the Garden of Eden itself, and access to the tree of life was denied (Gen.3:23-24).	to satisfy the greed of the wealthy and powerful, to the detriment of the poor and needy.[17]

Redeemed mankind has free access to both the city and the presence of God (Rev. 22:4, 14), and to the tree of life (Rev. 22:14), because ***the image of God in man has been fully restored.*** One of the effects of sin is the distortion of the image of God in man. This is a reflection of what one finds today in the many religions of the world and in their comprehension of God.

The book of Revelation tells us that God has a purpose for creation. John begins his gospel with this in mind:

"In the beginning was the Word, and the Word was with God, and the Word was God. He was in the beginning with God. All ***things came into being through Him, and apart from Him nothing came into being that has come into being"*** (John 1:1-3).

The world was created for the Son, by the Son, and through the Son. Many traditional churches recite in their liturgy, "In Him and with Him and through Him..." The world, in its pristine state, belonged to Him. Mankind was made to be His bride and was meant to enjoy creation with the Son of God. Adam and Eve did enjoy this intimate relationship with God before sin entered. Through sin, that which was created for the Son of God and His bride, was legally given to the devil, satan. He is the destroyer, and he reigns through death. However, the intrusion of sin and death can never permanently thwart or deter God's design and purposes. The time will come for God to remove the curse imposed upon creation, and bring to fruition His eternal purposes for creation.

The book of Revelation describes in detail the fulfillment of God's purposes. Its message tells us that satan, the "god of this world," cannot affect the earth's ultimate destiny, through his wicked

[17] *The Revelation Record* by Henry Morris, has a more in-depth study of the same spiritual truth. (Tyndale, 1983, pp 21-22).

plans for destruction. The end of all things is assured, because of Christ's obedience as man, His death on the cross, and His resurrection. His redemptive work has begun the renewal and restoration of all of creation. The glorious conclusion of His redemptive work is not merely the redemption of mankind; rather, the earth itself, even all creation will be redeemed and restored.[18] This will bring creation "back" to the original plan of God, but at a higher level of glory. **As God walked with Adam in the Garden of Eden, His Son will again walk with redeemed mankind in the renewed and restored earth.**

God gave us the prophetic word at the Arthur's Pass retreat: "I will walk with you in the snow as I walked with Adam in the garden. My Righteousness is like the great mountains. The snow is likened unto My garment of righteousness; I will walk with you in the snow as I walked with Adam in the garden. My righteousness will prevail upon you and upon the Nation, I AM covering you like a blanket of snow, *your iniquities and sin I will remember no more*. This is a day of atonement, even as you have drawn close to Me, My righteousness I have given freely to you. I see the desire in your hearts to worship Me, I will *open the doors of heaven* to My people to enter into worship before My throne. New understanding will be revealed of who I AM and the *glory of My presence* for you is seated in heavenly places with Me. I have called you a Nation a Holy Nation and a Royal Priesthood to minister before My throne. I have called, chosen, cleansed and sanctified you to worship before Me, casting your crowns in service to your King, as this is also the Father's will and the will of the elders before My throne. Even as I have covered you with a new garment of righteousness before Me, I am at work in your hearts to release you into a new realm, *the heavenly realm* and the desire of worship before Me. I am causing your hearts to be broken and to be stripped of the old that has hindered you, into

---

[18] For a more detailed theological discussion, see "*The restoration of the primordial world of Genesis 1-3* in *Revelation 21-22*." By Bruce Norman, PhD. 1997, *Journal of the Adventist Theological Society*, 8/1–2 (1997): 161–169. *www.bibelschule.info/.../Bruce-Norman—-The-Restoration-of-the-Primordial-World-of-Genesis-1-3-in-Revelation-21-22_25085.pdf*

free worship and preparing your hearts, so that *the new* may come forth as *Holiness unto the Lord.* A Royal Priesthood, accepted by the Father. Allow My purifying fire to come afresh into your hearts, which will set you free to worship in the Heavenly Realm, and to know, see and experience the throne of worship." Amen.

God has a Plan for all of His Creation

The book of Revelation talks about the coming Kingdom Age, which could be termed a righteous "new world order." However, satan is attempting to establish his wicked, counterfeit kingdom. Most bible scholars refer to the coming counterfeit government as a "one world government," or the "new world order." The Lord's awesome plan is to establish His righteous government over all the earth. This government will be based on the redemptive work of Christ. Satan's agenda is to enslave mankind through the human agency of a one-world government.

In His first coming, Christ redeemed the fallen nature of man, by "paying the purchase price" with His own blood on the cross. In His second coming, Christ will restore all of creation, by first stripping it of all that is distorted and contorted by sin. The spiritual world, the political world and the physical world will be overturned. The altars raised to false gods, the altars built to Mammon and man's financial systems, will be torn down. We have already seen this system of the "bulls and bears" crash in recent times.

Even in our modern world, which we think is so sophisticated, we have erected altars to Baal, Jezebel, and Moloch. You might wonder, "How could a modern society be worshiping these fallen powers?" Through the shedding of innocent blood, abortion, perversion, human trafficking, prostitution, and the sex slave trade, we empower these evil beings. They thrive on the blood of the innocent. Their detestable altars will be razed to the ground.

The internet pundits predict the crash of societal structures: economics, law and order, education, entertainment and culture (the seven mountains of culture, as some call it). We are beginning to see some of this taking place in nations around the world today. The

Lord Jesus will then subdue all things under His feet, as the author of Hebrews says:

"*You have put all things in subjection under his feet. For in subjecting all things to Him, He left nothing that is not subject to Him.* ***But now we do not yet see all things subjected to Him"*** (Heb. 2:8).

Christ will then rule and reign in equality and justice, and will eliminate the miseries and injustices that continually disfigure society, because of the presence of sin.

"*Therefore, since the children share in flesh and blood, He Himself likewise also partook of the same, that* ***through death He might render powerless him who had the power of death, that is, the devil,*** *and might free those who through fear of death were subject to slavery all their lives. For assuredly He does not give help to angels, but He gives help to the descendant of Abraham*" (Heb. 2:14-16).

The fact is that God has a plan for this world, and the church needs to join her Lord through worship, warfare and intercession, to hasten the coming of the Lord Jesus to the earth.

God Will Renew all of His Creation

It would be a very heartrending thought to think that God abandoned His creation to satan, allowing him to destroy it. First, it would concede defeat to the devil over God's creation. Secondly, it would render incomplete the redemptive work of the cross of Christ. God's plan is for the complete work of Christ's redemption on the cross: The redemption of man, and the restoration and renewal of the earth. All of sinful mankind's relationships with these fallen angels, has created an abode for these "gods" on earth. Most high places, in most religions across the world, claim the abode of these "gods." When I visited Mount Kailash in the Himalayas in India, I was enlightened to understand spiritual warfare and intercession for the high places. I found a pamphlet, which was left on the bus seat that took us up the mountain to Mount Kailash in the Himalayas.

It elucidated a belief that the high places on the earth have been claimed by these evil spirits:

"In the Rig Veda, the Himalayas are central to the worship of the cosmos. The mountain peaks are the petals of the Golden Lotus, which lord Vishnu created as a first step in the formation of the universe. On one of these peaks, Mount Kailash, sits Shiva in a state of perpetual meditation, generating the spiritual force that sustains the cosmos and all of life on the earth. He alone destroys, builds, plants, moves. All activity is guided by him."[19]

We find scriptures which highlight the spiritual importance of the high places on the earth.

The first example is seen in the temptations of Jesus by the devil:

"Again, ***the devil took Him to a very high mountain and showed Him all the kingdoms of the world and their glory****; and he said to Him, 'All these things I will give You, if You fall down and worship me'"* (Matthew 4:8-9).

We also see this phenomenon in the account of the transfiguration of Jesus:

"And Jesus was saying to them, 'Truly I say to you, there are some of those who are standing here who will not taste death until they see the kingdom of God after it has come with power.' Six days later, ***Jesus took with Him Peter and James and John, and brought them up on a high mountain by themselves. And He was transfigured before them; and His garments became radiant and exceedingly white****, as no launderer on earth can whiten them. Elijah appeared to them along with Moses; and they were talking with Jesus. Peter said to Jesus, 'Rabbi, it is good for us to be here; let us make three tabernacles, one for You, and one for Moses, and one for Elijah'"* (Mark 9:1-5).

---

[19] I found this pamphlet on a bus when travelling to Kailash with an intercessory team. There was no mention of the copyright of the pamphlet, so I am at a loss to give credit to the words I have quoted.

Another example is seen when the new Jerusalem comes down from heaven to earth:

"Then one of the seven angels who had the seven bowls full of the seven last plagues came and spoke with me, saying, ***'Come here, I will show you the bride, the wife of the Lamb.' And he carried me away in the Spirit to a great and high mountain, and showed me the holy city, Jerusalem, coming down out of heaven from God, having the glory of God. Her brilliance was like a very costly stone, as a stone of crystal-clear jasper.*** *It had a great and high wall, with twelve gates, and at the gates twelve angels; and names were written on them, which are the names of the twelve tribes of the sons of Israel"* (Rev. 21:9-12).

In every nation that I went to as part of a prayer team, we were explained that the high places were the abodes of the "gods" of the local religions. Christ will come triumphantly to crush these unlawful "tenants" to establish His kingdom, rule and reign.

"...And His voice shook the earth then, but now He has promised, saying, 'Yet once more I will shake not only the earth, but also the heaven.' This expression, 'Yet once more,' denotes the removing of those things which can be shaken, as of created things, so that those things which cannot be shaken may remain. Therefore, since we receive a kingdom which cannot be shaken, let us show gratitude, by which we may offer to God an acceptable service with reverence and awe; for our God is a consuming fire." (Heb. 12:26-29).

"Now it will come about that in the last days the mountain of the house of the Lord will be established as the chief of the mountains, and will be raised above the hills; and all the nations will stream to it. And many peoples will come and say, 'Come, let us go up to the mountain of the Lord, to the house of the God of Jacob; That He may teach us concerning His ways and that we may walk in His paths.' For the law will go forth from Zion And the word of the Lord from Jerusalem" (Isa. 2:2-3).

Whether the earth experiences earthquakes, tsunamis, financial collapse in many nations, or wars, God is overruling all things to work out His purposes. The church on earth needs to involve herself in these processes, through worship, warfare and intercession with the church in heaven, to "quicken," or as Peter says, "hasten" the return of the Lord.

"The Lord is not slow about His promise, as some count slowness, but is patient toward you, not wishing for any to perish but for all to come to repentance. But the day of the Lord will come like a thief, in which the heavens will pass away with a roar and the elements will be destroyed with intense heat, and the earth and its works will be burned up. Since all these things are to be destroyed in this way, what sort of people ought you to be in holy conduct and godliness, ***looking for and hastening the coming of the day of God,*** *because of which the heavens will be destroyed by burning, and the elements will melt with intense heat!* ***But according to His promise we are looking for new heavens and a new earth, in which righteousness dwells"*** (2 Peter 3:9-13).

CHAPTER 6

JOINING HEAVEN IN INTERCESSION

"When He had taken the book, the four living creatures and the twenty-four elders fell down before the Lamb, each one holding a harp and golden bowls full of incense, which are the prayers of the saints" (Rev. 5:8).

John sees the harps and bowls of intercessory prayer, offered by the elders in heaven. The Lord is inviting His church leadership to join the elders in heaven in worship, warfare and intercession. The most significant purpose of this intercession is to prepare the earth for the Lord's return.

In Part 1, we learned that as we participate in heaven's worship, the Bride is urging the Lamb to take the scroll, the title deed to the earth, and begins to open its seals. In the same way, as we align our intercession with heaven, we are crying out for the Lamb to begin the last redemptive judgments and preparations for the earth to receive and welcome her King. **He will not come to reign, apart from His Bride's desperate cry, welcome and invitation.**

The Lord is inviting us to purposefully structure our church worship, according to the pattern we have discussed already. As a result, this worship gives rise to intercession and warfare for the return of the Lord Jesus to the earth.

The Old Wineskins Will Not Work

Psalm 127:1 says, "*Unless the Lord builds the house, they labor in vain who build it.*" For many years, even centuries, sincere pastors and leaders have been conducting their worship services according to a familiar structure. We are moving into a new and critical time period, as this age comes to a close. The old patterns are not adequate to meet the intense spiritual demands of the coming of the Lord, and the battles that will surround His coming. Unless the church is aligned with heaven's ways, it cannot be an effective instrument in partnering with God's plan to restore and renew the earth. The Lord is building a spiritual house, made of living stones. He said, "I will build My church." For too long, man has been building His church, using building blocks of human reasoning, human desire, entertainment and convenience. It has been man-centered. The Lord said, "My house shall be called a house of prayer for all nations" (Isa. 56:7). We have done many things in church, but we have not been house a of prayer, worship, or intercession. How can He return to earth, until He hears the fervent cries of His Bride?

Church leadership is responsible for the structure of church services. The pastors of the church on earth are expected by heaven, to join with the elders (in heaven), in their heavenly worship. This will mean casting your crowns before the Lord of heaven and earth. This will mean patterning your corporate worship in a way that joins with heaven's worship and intercession. It means a new makeover of church services, heaven centered, and patterned.

The Role of the Holy Spirit in the End Times

The Holy Spirit was instrumental in the primordial creation of the earth.

"*In the beginning God created the heavens and the earth. The earth was formless and void, and darkness was over the surface of the deep,* ***and the Spirit of God was moving over the surface of the waters"*** (Gen. 1:1-2).

Psalm 104 gives a prophetic description of the Lord moving in power, as He created the heavens and the earth.

"*Bless the Lord, O my soul! O Lord my God, You are very great; You are clothed with splendor and majesty, covering Yourself with light as with a cloak, stretching out heaven like a tent curtain. He lays the beams of His upper chambers in the waters; He makes the clouds His chariot; He walks upon the wings of the wind; He makes the winds His messengers, flaming fire His ministers. He established the earth upon its foundations, so that it will not totter forever and ever*" (Psa. 104:1-5).

In these last days, the Holy Spirit is working through the intercessions of His Bride, to renew the earth.

"*You send forth Your Spirit, they are created; and You* ***renew the face of the ground***" (Psa. 104:30). (The word "ground" would be better translated as the "earth.").

We are aware of the Spirit's works of restoration, healing and deliverance within us. But now, we must also become very aware of the larger redemptive purposes of the entire creation. The Lord is so redemptive in nature, that He will not merely restore things as they were in the beginning, but He will recreate them at a higher level of glory in end.

Even as the Father, Son and Holy Spirit created all things in the beginning, so the fullness of the Godhead ("The Trinity") is active in this last days' ultimate regeneration of all creation.

God has given His Holy Spirit to His people, so that they will do their part in the work of the renewal and restoration of the earth.

The First Judgment – The Partial "Wrath" of the Lamb

"*And I saw when the Lamb opened one of the seven seals, and I heard one of the four living creatures saying as with a voice of thunder, 'Come.'*

"And I saw, and behold, a white horse, and he that sat thereon had a bow; and there was given unto him a crown: and he came forth conquering, and to conquer. And when he opened the second seal, I heard the second living creature saying, 'Come.'

"And another horse came forth, a red horse: and to him that sat thereon it was given to take peace from the earth, and that they should slay one another: and there was given unto him a great sword. And when he opened the third seal, I heard the third living creature saying, 'Come.'

"And I saw, and behold, a black horse; and he that sat thereon had a balance in his hand. And I heard as it were a voice in the midst of the four living creatures saying, A measure of wheat for a shilling, and three measures of barley for a shilling; and the oil and the wine hurt thou not. And when he opened the fourth seal, I heard the voice of the fourth living creature saying, 'Come.'

"And I saw, and behold, a pale horse: and he that sat upon him, his name was Death; and Hades followed with him. And there was given unto them authority over the fourth part of the earth, to kill with sword, and with famine, and with death, and by the wild beasts of the earth" (Rev. 6:1-8).

Biblical Chaos is an Important Understanding for Intercession

In the book of Revelation, John introduces the processes of change on the earth, which begin with the Lamb breaking the seven-sealed scroll. In biblical times, important documents were sealed, often with seven seals.

As the Lamb breaks open the first four seals, John saw the four horsemen described above. These horsemen are called into the presence of the God by the mighty angels, the Cherubim. They cannot be evil spirits, as John saw them in heaven. Nothing profane, not one iota of sin will be entertained in the presence of God. Hence, they are not evil spirits, but as Psalm 104:4 says that they are "His messengers, ministers of flaming fire." They do not come to do their own will. They are angelic beings of mighty power (see Rev. 7:1).

Throughout scripture, we have been given examples of their power to affect the earth: Sodom and Gomorrah are vivid examples. Then again, we have the account in 2 Kings 19:35, where just one angel slew 185,000 men in the Assyrian camp. The four horsemen strip the earth of its provision for the nations and for mankind, based on how each nation has responded and lived according to the Word of God. Their righteous judgments hurl the earth into a condition of "chaos." It is interesting that "chaos" is the bedrock upon which restoration and renewal are brought forth by the Holy Spirit. The Lord has always worked to bring order out of chaos.

In Genesis 1, the Holy Spirit hovered over a chaotic world, to bring it to its perfection, its pristine glory by the declared word of God. The pre-Genesis (chapter 1 and 2) condition of the world was a chaotic world. The prophet Isaiah, in his prophecy about the fallen angels, reveals this to us, that when they were cast out of heaven, they were sent to the earth.

"*How you have fallen from heaven, O star of the morning, son of the dawn!* ***You have been cut down to the earth,*** *you who have weakened the nations! But you said in your heart, 'I will ascend to heaven; I will raise my throne above the stars of God, and I will sit on the mount of assembly In the recesses of the north. I will ascend above the heights of the clouds; I will make myself like the Most High.' Nevertheless you will be thrust down to Sheol, to the recesses of the pit. Those who see you will gaze at you, they will ponder over you, saying, 'Is this the man who made the earth tremble, who shook kingdoms…*" (Isa. 14:12-16).

Adam and Eve, in disobeying God, unleashed yet another chaos into the created order of the pristine earth, bringing death and decay into every sphere of man's life. **Into the corroded and chaotic world caused by man's sin, the Holy Spirit once again "hovers" over Mary, to bring forth the Messiah Yeshua, the Lord Jesus, into our chaotic world.** By His death on the cross, He drew out a people from the domain of chaos, sin and death, into the domain of the Son. In Him, we enter the secret place of rebirth, regeneration, restoration and peace in the midst of our chaotic world. Therefore, the plan of redemption was partially fulfilled in the birth of the church, which occurred with the first coming of our Lord Jesus. He is coming again, back to a world which will be thrown into even worse chaos, before all things are renewed.

For us to intercede with the heart and mind of Christ, we must understand heaven's pattern.

- The four horsemen, the seven trumpets and the seven bowls, are judgments sent from the throne room of God, which create chaos on the earth.
- The unity of worship, warfare and intercession of heaven and earth, in the power of the Holy Spirit, brings forth the second coming of Messiah, for the redemption of the physical earth. He comes as the Lord of all lords and the King of all kings, and sets up His kingdom of everlasting peace, joy and justice.

If we cannot see that the chaos we are now experiencing, which will get even worse over the coming years, is coming from the throne room of God, we will find ourselves praying against the purposes of God. God forbid that this should happen in the church. It will be prayer in vain.

There will be times when the calamities around us will indeed be the righteous, redemptive judgments of God. In these cases, as we discern rightly, we should not rebuke the Lord's angelic servants from their assignments. However, there will also be cases when fallen angelic powers will create destruction and chaos, which the "Sons of the Kingdom" MUST rebuke and resist.

The term "sons of the kingdom," refers not to the church at large. It describes a company of believers who have paid the price to step into their inheritance, while still on this earth (see Rom. 8:19). This inheritance, which is available to all believers, includes the restoration of the authority of the spoken word, which Adam possessed before he fell. Many respected prophets have seen this company in prophetic vision. As Daniel, they will do exploits, and will shine like the sun in the kingdom of their Father (see Dan. 11:32, 12:3).

These are sons of the Kingdom. These exploits will include standing against hurricanes, tidal waves and storms. The winds and waves obey the creative spoken Word of God, as demonstrated by our Lord Jesus when he walked the earth during His first coming. Similarly, even if the black horse rider brings famine on the earth, the sons of the kingdom will see the multiplication of food, as did our Lord Jesus. Their bread and their water are sure (see Isa. 33:14-16).

Before we try to rebuke spirits that are causing chaos in the nations, we should discern whether this is a judgment of God or the cruelty of satan. We do not want to find ourselves fighting the wrong spiritual battle, and may find ourselves in the enemy's camp rather than in the Lord's camp. This happened to Peter, who was shocked when the Lord Jesus said He must go to the cross. He wanted to stop the Lord Jesus from making that decision, because the cross was not in his paradigm. It is a very humane thing to do, no doubt, but it is the temptation of the flesh and a misunderstanding of heaven's ways. The word of God must influence our paradigm of the ways of God. The Spirit of the Lord will help us discern what comes from the throne room of God and what emanates from the pits of hell. The church must stand against hell's destructive ways, yet bow to heaven's judgments, even if it means pain and suffering will follow.

Intercession with heaven calls for the greatly encouraging word that Joshua received, when the Lord commissioned him to take the promised land: "Only be strong and of good courage." Events that will unfold before our eyes in the days ahead, will cause us to tremble and fear, but "be strong and of good courage." The Lord Jesus added this word, to that command to Joshua: *"I have overcome the world."* **For our worship to be authentic and our intercession effective, the only choice is to join with heaven.**

The Stripping of the Earth Begins

Let us try to understand who these four horsemen are, what they symbolize, and what their purpose is. Why they are sent out from the presence of God in heaven? With this understanding, we can enter into intercession as it is in heaven, and with the saints and angels.

*"Now I lifted up my eyes again and looked, and behold, four chariots were coming forth from between the two mountains; and the mountains were bronze mountains. With the first chariot were red horses, with the second chariot black horses, with the third chariot white horses, and with the fourth chariot strong dappled horses. Then I spoke and said to the angel who was speaking with me, 'What are these, my Lord?' The angel replied to me, **'These are the four spirits of heaven, going forth after standing before the Lord of all the earth.'**"* (Zech. 6:1-5).

Zechariah's verses above tell us that the "four horsemen" are Revelation's "four spirits of heaven." The four angelic beings come from the presence of God in heaven Revelation 7 reveals their identity.

*"After this I saw **four angels** standing at the four corners of the earth, holding the four winds of the earth, that no wind should blow on the earth, or on the sea, or upon any tree"* (Rev. 7:1).

Psalm 104 tells us that they are God's messengers, ministers of fire. If you remember, the Cherub with a flaming fiery sword stood guard at the entrance of the Garden of Eden, after Adam and Eve were banished from the garden, so no one would enter and eat of the fruit of the tree of knowledge.

"He makes the winds His messengers, flaming fire His ministers" (Psa. 104:4).

We know from the study of earth sciences that there are four winds called "jet streams." These jet streams are responsible for bringing about seasonal changes to the earth. Scripture is telling us that the four horsemen are powerful angelic beings that control these jet streams to do the will of God. A news article on December 22, 2010, talks about how the jet stream affects weather patterns. The article said that a strange change in direction of the jet stream brought snow in Australia's summer season, and warm temperatures in Greenland's winter. [20]

Revelation 6 tells us that the release of the four horsemen, puts into effect the stripping of the earth at four different levels. This will bring hardship and chaos to the earth.

Today, no matter which nation we come from, in some measure, we experience the economic crisis that the entire world is experiencing. There are many reasons propounded why this economic crisis happened, and some predicted it accurately. However, the real reason for the crisis is not very well known. The cause is likely to be more of a spiritual factor, than that of a natural economic phenomenon. Yes, the economic forces of excessive spending, national debt, natural disasters and the like have contributed their share to the crisis. At the core of the matter, there is a spiritual aspect that the scriptures reveal, which unfortunately, world leaders will ignore, rather than acknowledge that God has the answer.

In every age of human history, the four stages of a society's downfall, are represented by the four horsemen. We see for example, these stages in the downfall of the following empires: the Greek, the Roman, the Spanish, and the British. In our modern society, the process of societal collapse remains the same. Today, however, we are moving into the ultimate pattern in the revelation of God's word. In these last days, we will see the stripping of the old heavens and the old earth, and the coming of the new heavens and the new earth.

The White Horse Rider

"*I looked, and behold, a white horse, and he who sat on it had a bow; and a crown was given to him, and he went out conquering and to conquer*" (Rev. 6:2).

The white horse rider, according to the revelation I received in the vision of heavenly worship and intercession, represents the Word of God. The Word of God has gone out to the ends of the earth,

[20] To read this article in detail please check Mail Online web site. http://www.dailymail.co.uk/sciencetech/article-1340436/Why-cold-warm-Greenland-Diverted-jet-stream-letting-icy-blast-Arctic.htm

conquering and to conquer the hearts of mankind. It beckons man to return to God through repentance and acceptance of Jesus Christ as Lord and Savior. John's opening remarks in his gospel, presents the Lord Jesus as the Word of God. In the beginning, the Word of God existed and the Word was God. Wherever man is confronted with the gospel, he is confronted with the Lord Jesus Christ, the Word of God, knocking at the door of his heart. (Rev. 3:20).

Irenaeus was an influential Christian theologian of the second century. He was among the first to interpret this horseman as Christ Himself, His white horse representing the successful spread of the gospel. Various scholars have since supported this theory, citing the later appearance, in Revelation 19, of Christ mounted on a white horse, appearing as the Word of God.[21]

My understanding of the role of the rider of the white horse is in line with some of the well-accepted commentaries.

Clarke's Commentary on the Bible

"**Rev 6:1: When the Lamb opened one of the seals**: It is worthy of remark that the opening of the seals is not merely a declaration of what God will do, but is the exhibition of a purpose then accomplished; for whenever the seal is opened, the sentence appears to be executed.

Rev 6:2: A white horse: It represents the gospel, and pointing out its excellence, swiftness, and purity.

He that sat on him: Is supposed to represent the Lord Jesus Christ.

A bow: (Represents) the preaching of the Gospel, which causes a darting conviction into the hearts of sinners to repent and receive salvation.

A crown: Which is the emblem of the kingdom of God which Christ is to establish on earth.

---

[21] "Four Horsemen of the Apocalypse" From Wikipedia, the free encyclopedia. http://en.wikipedia.org/wiki/Four_Horsemen_of_the_Apocalypse

Conquering, and to conquer: Overcoming and confounding the Jews first, and then the Gentiles; spreading more and more the doctrine and influence of the cross over the face of the earth."[22]

Matthew Henry's Concise Commentary: The Rider of the White Horse

"**Rev 6:1-8** Christ, the Lamb, opens the first seal: a rider on a white horse appeared. (White horses are generally, not used in war, because they make the rider a mark for the enemy). By the going forth of this white horse, a time of peace, or the early progress of the Christian religion, seems to be intended; it is going forth in purity, at the time when its heavenly Founder sent his apostles to teach all nations, adding, lo! I am with you always, even to the end of the world. The Divine religion goes out crowned, having the Divine favor resting upon it, armed spiritually against its foes, and destined to be victorious in the end."[23]

A Coming Famine of the Word of God

We live in an age where the ultimate fulfillment of God's prophetic word is unfolding before our eyes. The Rider of the White Horse is reaching the end of the road. **Rejection of God's saving word will result in judgment upon the nations.** This judgment on the nations will come through the riders on the red horse, the black horse and the pale horse. The rider of the white horse will not ride this path again. Each generation, and each church has received the word of God sent to them. Heaven expects a response to the word of God. "*I know your deeds, that you are neither cold nor hot; I wish*

---

[22] Above Commentary is from: http://www.sacred-texts.com/bib/cmt/clarke/rev006.htm, 1831, public domain

[23] Above Commentaries is from: http://www.sacred-texts.com/bib/cmt/mhcc/rev006.htm, 1706, public domain.
The following remark is on this web site: "Public domain files from this site can be used for *any* purposes. You may… Quote them in part or entirety in print: in a book, magazine, etc. without asking anyone (including sacred-texts) for permission." See http://www.sacred-texts.com/cnote.htm for the above remark.

that you were cold or hot. So because you are lukewarm, and neither hot nor cold, I will spit you out of My mouth." (Rev. 3:15-16).

The angel of the white horse is looking for the fruit of repentance and the return to the Lord for salvation. This is the last opportunity for mankind. We see this in the book of Revelation, as well as in the words of the Lord Jesus in Matthew 24:14: "A*nd then shall the end come.*" Mankind does not have the opportunity for another chance, for time has run out. The prophet Amos prophesied about the days of famine of the word of God.

"*Behold, days are coming, declares the Lord God, when I will send a famine on the land, not a famine for bread or a thirst for water, but rather for hearing the words of the Lord. People will stagger from sea to sea and from the north even to the east; they will go to and fro to seek the word of the Lord, but they will not find it. In that day the beautiful virgins and the young men will faint from thirst*" (Amos 8:11-12).

The first horseman from heaven on the white horse brings hope to mankind. Has mankind accepted or rejected God's word? Have the nation that have received the gospel lived their lives according to God's word? The white horse rider tells us to "make haste," make haste to repent, make haste to reconcile and return to God. It is the period of grace for all mankind.

"As it is written in the book of the words of Isaiah the prophet:
The voice of one crying in the wilderness, make ye ready the way of the Lord.
Make his paths straight. Every valley shall be filled, and every mountain and hill shall be brought low. And the crooked shall become straight, and the rough ways smooth; and all flesh shall see the salvation of God." (Luke 3:4-6).

It is also a time for the bride to get herself ready, to live by the word of God.

How can we Intercede?

The book of Genesis tells us that God has always been at man's side even when he rejected God. This is so evident to us, as we hear thousands of testimonies of people who gave their lives to God. Yes, all flesh is given the opportunity to hear the word of salvation in the Lord Jesus Christ alone, to respond or to reject that invitation. This is the moment of truth. The release of the white horse over the earth gives us hope that the famine of God's word has not yet begun. Heavens gates are still open to repentant hearts. The Father of light is still at the gates, looking for His children to come in, before the gates are closed. The Lord is making available the grace of God for one last opportunity. Matthew 24:14 says, "*And this gospel of the kingdom will be preached in the whole world for a testimony unto all the nations; and then shall the end come.*"

We still have an opportunity for the church on earth to join heaven in prayer and intercession and warfare for all our loved ones who are resisting the word of God and its power to save. It is also an opportunity to pray for our neighborhood, our cities, our nation and the nations of the world. The church needs prayer too! Wherever the church has "lost her way," the body of Christ must pray that she gets back on track. There are denominations or branches of the church, which have mingled other religious doctrines with solid biblical doctrine. This is called "syncretism". It is more deadly to the true life of the church than blatant falsehood. By mingling the names and practices of other gods and other principles of life, the gospel is diluted, polluted, distorted rendering it ineffective for the saving of souls. One example of this is that a certain branch of the Christian church conducts "Buddhist teaching retreats" for its Christian followers! They feel that they can mingle portions of Buddhist teaching with Christianity, and have a better package. This will never bring a soul to salvation, because the word of God is nullified by an unholy mixture.

It is a time for supplication and intercession, to wage warfare on the "Pharaohs," the evil spirits (see Ephesians 6) that holds mankind bound to sin, and to say to them, "Let my people go."

A Simple Model for Intercession

Notice several things in this sample prayer below. First of all, it shows us how to petition the Lord to send the rider of the white horse to various cities or regions, where the word must be heard. It speaks to gates of resistance, and commands door to open where the word is resisted.

More importantly, please notice this warning: We do not bind and rebuke powerful principalities and rulers in dark places, who reign over regions with wickedness. Even Michael, the Archangel did not rebuke satan directly, as we learn in Jude, but he said, "The Lord rebuke you, satan!" If this mighty being did not rebuke satan directly, we should humbly follow his pattern. Notice below, the prayer warrior says, "Lord, bind the Prince of Persia." Whenever we do warfare, please cover yourself and your loved ones and all that which is under your sphere of authority with THE BLOOD OF JESUS, before or after doing this warfare. Speak it aloud, for it is mighty to protect you and your family. It is not a magic formula. The blood is living force, and you must understand the power of the blood, or saying the words will not help you.

"O Lord, send Your white horse and its rider to our nations with the gospel of life, so that people still have an opportunity in this period of grace to receive eternal life in Christ Jesus. Lord, send Your rider of the white horse to the nations that have fettered the preaching of the gospel (Here, the church can focus prayer on particular nations), to force open the doors to the gospel, for example: Lord, bind the prince of Persia that still grasps the freedom of the people of those nations. In Your name, we command these gates in the heavenly realms, "be opened in the mighty name of the Lord Jesus," and make way for the rider of the white horse, all over the regions of Persia, Asia and wherever the gospel experiences resistance through political power and regimes. Lord, You open the doors with Your mighty roar over these nations. Thank You, Lord."

The Second Stage: The Red Horse Rider

"*When He broke the second seal, I heard the second living creature saying, 'Come.' And another, a red horse, went out; and to him who sat on it, it was granted to take peace from the earth, and that men would slay one another; and a great sword was given to him*" (Rev. 6:3-4).

The rider of the white horse brought the gospel of God's saving grace to the nations. When a nation rejects the opportunity of God's free gift of eternal life through His Son, that nation opens itself to God's wrath. This may be a harsh word, and we may find some theology to get around this declaration. Whichever way we may want to circumvent the truth, the hard facts is that nations will be accountable for the rejection of the gospel, and they will face the wrath of God.

"*He who believes in the Son has eternal life; but he who does not obey the Son will not see life, but the wrath of God abides on him*" (John 3:36).

"*For God so loved the world, that He gave His only begotten Son, that whoever believes in Him shall not perish, but have eternal life. For God did not send the Son into the world to judge the world, but that the world might be saved through Him.* ***He who believes in Him is not judged; he who does not believe has been judged already,*** *because he has not believed in the name of the only begotten Son of God. This is the judgment, that the Light has come into the world, and men loved the darkness rather than the Light, for their deeds were evil*" (John 3:16-19).

The red horseman's task is to take peace from the earth. This seems so true in many parts of the world today. Some writers believe that the color red meant the red horse rider was the spirit of communism. However, this may have been true in some past "age" of the world, when communism brought a lot of terror among nations. Currently, I think it means, any nation that does not experience peace, whether through communism, or fundamental islamization, or any other means of forced domination, there the red horse rider is very active. The word peace in the bible is more than just the cessation from hostility or war. The Hebrew meaning of the word "shalom" has a very different meaning. The Old Testament Hebrew Lexicon (Strong's Number 7999) gives us a deeper understanding:

1. to be in a covenant of peace, be at peace
    1. to be at peace
2. to be complete, be sound
    1. to be complete, be finished, be ended
    2. to make whole or good, restore, make compensation
    3. to make an end of [24]

In summary, the verb form of the root word is '*shalam*,' which is usually used in the context of making restitution. The verb '*shalam*' literally means to make whole or complete. For example, the peace that passes all understanding is peace that comes from heaven. This brings wholeness in relationships. Peace in the Hebrew is a relational word. It is peace with God and with one another. In the English language, we use it in a non-relational sense. A common expression that most of us are familiar with is, "I am at peace about this situation." What that really means is that one has accepted the outcome of that situation. It has nothing to do with personal relationships. When the scriptures talks about peace, it means a personal relationship between God and man, and between people in a nation.

The rider of the red horse comes to take peace from the nations that have rejected the gospel previously brought by the rider of the white horse. Those who are not at peace with the one true God, will not be at peace with themselves nor with one another. Even the fundamental peace within a family, which binds the family in filial love, is taken away. The family unit is the core of society, and as goes the family, so goes the society. We will see in cities and in the nations,

[24] *The Old Testament Hebrew Lexicon is Brown, Driver, Briggs, Gesenius Lexicon; this is keyed to the "Theological Word Book of the Old Testament." These files are considered public domain.* http://www.searchgodsword.org/lex/heb/view.cgi?number=07999

riots of all kinds. Lawlessness will increase, and it will be common to hear of children revolting against their parents and "sibling wars" leading to murder, rape and death. Peace, the foundation of civilized society is breaking down. Our newspapers are full of this lately.

If nations and governments have not made their peace with God, how can treaties bring lasting peace between warring nations? Efforts past, present or even future to bring peace to the Middle East by the U.N., or the US, or Europe have failed. If the fundamental spiritual problem of peace with the one true God and His Son, the Lord Jesus Christ is not restored, peace between nations will either be fragile or nonexistent.

When we read Revelation 6:4, we immediately think of words of the Lord Jesus.

"*And you shall hear of wars and rumors of wars, see that you be not troubled: for these things must come to pass but the end is not yet. For nations shall rise against nations, and kingdoms against kingdoms; and there shall be famines and earthquakes in different places*" (Matt. 24:6-7).

When the gospel brought by the white horse is rejected, what will follow will be the red horse of judgment. He will trigger heaven's partial judgment that is in his mandate to take peace away from the nations. Earthquakes and famines, wars and upheavals, revolts and riots, will be the result of rejecting God, something that we will see on the increase in the coming days and years.

Intercession for the Nations: In Wrath Remember Mercy

How does the red horseman beckon us to intercession? How can the church stir itself to pray for these times without fainting at the very enormity and severity of the judgment that each horseman brings? We cannot come against God's righteous judgments over the nations. But we can remember one thing as we pattern our worship as is in heaven, God's mercy seat is set up in that place of prayer. An open portal to heaven created for the church to plead for God's mercy for the nations. In the heavenly worship, the elders lift up their bowls of incense and intercession to Him, so that He might show His mercy on the nations. Will the elders of the church join them likewise, at every church service pleading for God's mercy on their cities and nation all over the world, wherever the church exists? Is our intercession for the nations prayers like breadcrumbs that fall off the altar of communion? It is hard to fathom the depth of God's love and mercy and compassion. In the midst of judgment, He can still show mercy and compassion, and show favor to the nations we pray for, in accordance with His revealed word. He waits for His bride to believe in Him and to plead with the only one who can bring mercy in a time of judgment. Can He get His bride's attention to this?

Recently, in Eastern Australia, more people died in severe floods than when a cyclone of the proportions of hurricane Katrina[25] struck the cities. The difference was that not a single person died in the cyclone, because prayer went up. Christians cried their hearts out to God everywhere, in shelters and in churches. The little group in New Zealand, where my wife and I share our Christian life, met for special prayer for Australia. Everywhere, on bended knees and heads bowed low, people cried out to the Lord: "No Lord, not again, have mercy, spare your people, spare Australia, we plead O Lord."

God's mercy came down. The gigantic cyclone deflected just enough, and on landfall missed the larger cities. Its roaring ferocity reduced, and its damage lessened, which could have been severe, had it hit the major city (Cairns) directly. The red horse rider is visiting Australia. Western Australia was facing firestorms, which were burning up houses and destroying property. Eastern Australia was facing incessant rains and cyclones in the same week. Every nation on the earth will receive a visit, and woe to that nation weighed in God's balance and found wanting by the standards of God's word.

---

[25] Visit Fox News for some excellent videos of this devastation storm that hit the US 5 years ago and changed the coastline forever. http://www.foxnews.com/story/0,2933,168132,00.html and http://www.foxnews.com/scitech/2010/08/28/katrina-years-coastline-changed-forever/

"Lord, I have heard the report about You and I fear. O Lord, revive Your work in the midst of the years, in the midst of the years make it known; ***in wrath remember mercy****"* (Hab. 3:2).

This is why prophecy is so important. Prophets need to hear the heartbeat of God and pray, "Lord in your righteous judgment, remember mercy." I remember a responsive song that is sung in some traditional churches even today: "Let you mercy be on us, O God, as we place our trust in You." When we see nations like China and India being blessed with God's favor, we ask why? It is only because the persecuted church of those nations continues to intercede for God's favor and blessing. God comes to show mercy in response to His children's cry. China and India have received warnings from God through famine, floods, and earthquakes. If these nations continue to reject the word of God, then in God's time, the red horse rider of God's judgment will not fail to pay a visit.

Does the church in New Zealand intercede with the same fervor as the underground church in China intercedes for her nation? The churches in every nation must examine themselves to see if they are either hot or cold. For if they are lukewarm in this hour of intercession, they have strayed from God's will for them to be a "house of prayer for all nations" (Isa. 56:7, Matt. 21:13). We need to pray that our leaders will respond to the pain and suffering of God's righteous judgments by humbling themselves and returning to God in repentance and reconciliation.

Intercession for Israel: Why pray and intercede for Israel?

There is a blessing for a nation when it prays for the peace of Jerusalem. New Zealand has a great heritage in that blessing, when in W.W.I; its soldiers laid down their lives by taking Jerusalem out of the hands of the Turks. Since New Zealand was fighting under the British command, Jerusalem came into British hands. The Balfour Declaration of Great Britain paved the way for Israel's independence as a nation, accepted and declared by the U.N. in 1948. Scriptures indicate that the prosperity of a nation is tied to praying for the peace of Jerusalem, with a heart of love for the nation of Israel:

*"**Pray for the peace of Jerusalem: May they prosper** who love you"* (Psalm 122:6).

This is a scriptural pathway to prosperity, not merely in the material sense, but as God's favor, provision and protection, an embracing, abiding assurance of His love to those who obey His word.

Imagine if your great grandfather buried a big treasure under the foundations of your house. What would you do? You could say that this was a good fairy tale and forget about it. If you are like most people, the thought that there is a treasure just under your nose, would be haunting you every day. Then one day, maybe you or someone in your lineage would want to deal with the haunting dream. That person will tear down the house to find the treasure. So it is with every nation. Those who read the scriptures can act on this word and leave a blessing for all posterity in the nation. Alternatively, they can deny their children that blessing, hidden in the promise of God's word.

Pray for the peace of Jerusalem with a heart of love. God will then open His treasure for that nation.

The Gentile Church is Indebted to Israel for her Salvation.

The Lord Jesus told the Samaritan woman at the well that He was a Jew, and He lived and worshiped as a Jew.

"*You worship what you do not know; we worship what we know, for salvation is from the Jews*" (John 4:22).

Very recently, the Greek Orthodox archbishop of Massachusetts, Cyril Salim Butros made the statement that Christ nullified God's promises to the Jews.[26]

---

[26] For a full article on this news see Israel Today Magazine October 24, 2010. http://www.israeltoday.co.il/default.aspx?tabid=178&nid=22093

"The Holy Scriptures cannot be used to justify the return of Jews to Israel and the displacement of the Palestinians, to justify the occupation by Israel of Palestinian lands."

This would mean that the "promised land," which is Israel today, is not justified. It means blotting out a nation that came into being by God's sovereign will and by a U.N. resolution.

He then escalated the situation by declaring that the original promises made by God to the children of Israel *"were nullified by Christ. There is no longer a chosen people."* The Greek Orthodox archbishop, Mr. Butros, rejected Israel as "the Jewish state."

I do not wish to sit in judgment over a professing Christian. However, when heads of churches or nations make statements or act against Israel, God's holds them and their church accountable. This makes the burden of sin all the weightier. In rejecting Israel as God's chosen people, there presents a danger in stepping out of the cover of God's protection. The Lord said to Abraham, *"I will bless those who bless you, and curse those who curse you."* This indeed is a warning. For the curse of God, like the wrath of God is something every man, church or nation must fear.

The disappearance of the seven churches in the book of Revelation, can happen to any church. A church has lost its "saltiness," when it denies the scriptures. In denying the scriptures, they are actually denying the Lord Jesus Christ, who is the Word of God. The scriptures become the very word that judges us:

"*You are the salt of the earth; but if the salt has become tasteless, how can it be made salty again?* ***It is no longer good for anything, except to be thrown out and trampled under foot by men"*** (Matt. 5:13).

Why did God choose Israel, rather than another nation, which was more desirable or powerful? Why did He rescue and redeem the pitiful Hebrew slaves, who labored in Egypt's sweltering mud pits under a tyrannical Pharaoh? Because God is just. He works with that which is insignificant and weak to confound the strong and wise (see 1 Cor.1:20-25).

We see this pattern in the coming of the Lord Jesus Christ.

- In a time of chaos and suffering in Egypt, Israel came forth by Yahweh's hand of deliverance.
- In a time of chaos and suffering in Israel, ruled by Herod, another harsh "Pharaoh," type, the Lord Jesus came forth. This was His first coming.
- In a time of chaos and suffering in the earth, the Messiah will come again, to reign and rule as king over all the earth.

To ignore Israel is to ignore one part of God plan of salvation, one half of the Lord Jesus' bride, God's chosen people, the apple of God's eye. Both the redeemed descendants of Sarah and the redeemed descendants of Hagar (the Jewish people and the gentile people), are purchased by the blood of the Lord Yeshua (Jesus) on the cross.

Israel and the nations around her desperately need the Messiah's peace. The church must pray for Israel and the nations of the Middle East **prophetically**, according to His end-time purposes, which the word of God declares. God's desires salvation for all men, and therefore, all nations, to come to the knowledge of His truth (see 1Tim.2:4). The Holy Spirit will guide the church in her intercession in praying according to God's plans, God's desire, and not human desire or design.

The past and present relationship between the church and the Jewish people is scarred by events both political and religious. From the second and third century down to the present, there is need for deep healing of these relationships. Intercessors need to study the historical past and pray for healing so that the end-time, "one new man" church comes forth in power. This church needs to fulfill her end-time calling. For an in depth treatment of this vital biblical truth, I recommend the reading of Jill Shannon's book, "A Prophetic Calendar," (Destiny Image Publishers, 2009), and her forthcoming book on Israel.[27]

---

[27] "Israel's Prophetic Destiny: If I Forget Jerusalem." Destiny Image Publishers, 2011. Check for this book on Jill's web site www.coffeetalkswithmessiah.com

A Prayer Example: When Peace, by the Red Horseman, is taken Away

"God, even as You enforce Your judgments through war, may Your mercy open the minds and hearts of politicians and parliamentarians, army generals and religious leaders, to see the light of Christ. Lord, even as your strip the nations for the restoration and renewal of the earth, may it bring forth many sons and daughters in Christ."

"May true peace come forth through the bowing of the head and the bending of the knee to Jesus Christ, Prince of Peace, the only Lord and Savior of the nations. Lord, we pray for Your church, Holy Spirit, guide the leaders in bringing Your people into their coming role of reigning and ruling with you in righteousness, peace and joy, when you return to the earth as King of all kings and Lord of all Lords. Let this permeate every sphere of life that You have called us to live in. May Your Kingdom come, may Your will be done on earth as it is in heaven.

"Father, we pray that our nation will respond to your word and warnings before the Day of Judgment. Lord, may Your rider of the red horse bring down false structures of slavery to false gods and tear down the altars raised to them. May the eyes of the people will be opened and allow them to see falsehood where it exists, and grace in abundance to choose to follow You. In doing so, may they leave a blessing in that place, for where true biblical peace exists, there nations will dwell in unity, oneness, peace and joy. Amen"

Intercession: The Church as Ambassadors of God's peace

The earth is being prepared for the marriage feast of the Lamb. The bride has a role in this marriage preparation. Jesus' words in a parable in Matt. 22:11-12 indicate this truth:

"*But when the king came in to look over the dinner guests, He saw a man there who was not dressed in wedding clothes, and He said to him, 'Friend, how did you come in here without wedding clothes?' And the man was speechless.*"

The bride's preparation is not cosmetic by nature. She is to be an ambassador of the peace of God on earth. She is to broker peace with God and man through intercession, and as an ambassador of reconciliation.

"*Therefore,* ***we are ambassadors for Christ****, as though God were making an appeal through us; we beg you on behalf of Christ, be reconciled to God.*" (2Cor.5:20).

This will require the church to disciple her flock to be ambassadors of peace at the various levels of society in which they live. From those in high offices of government, to simple jobs in day-to-day life, all members of body of Christ should live out their role as an ambassador of Christ's peace to the world around them.

Intercession for the Church: Disciple the Nations

"*And Jesus came up and spoke to them, saying, all authority has been given to Me in heaven and on earth. Go therefore, and make disciples of all the nations, baptizing them in the name of the Father and the Son and the Holy Spirit, teaching them to observe all that I commanded you; and lo, I am with you always, even to the end of the age*" (Matt. 28:18-20).

Nations lost their peace with God when governments separated church from state. They exchanged the wisdom of God for the folly of fallen man. Because of this, the red horseman rides on in every age, wherever man rests on his pride to "do it without God." In this final age before the kingdom of God comes to the earth, God will cause the fall of nations. Therefore, the church has been given the mandate to disciple the nations now, so that the mature Bride will be equipped to rule and reign with Him when He comes. The church needs to understand that we are not merely discipling people so that they can go to heaven when they die. We are preparing them for governmental authority learning to reign and rule with the Lord Jesus on the earth.

The second coming of the Lord brings into focus an important aspect of the discipleship of the nations. However, we would first need to understand the concept of the rapture. To understand the rapture, we must first look at a commonly used Greek word from the New Testament: "parousia." This word is used twenty-four times in the New Testament, to mean the "coming, arrival, or visitation." Most often, it indicates the second coming of the Lord Jesus Christ to the earth.

Another Greek word often taken to mean the "rapture" is "*apantesin.*" This word is used less often in the New Testament, but wherever it is used, it means to "receive and escort an important person." In the parable of the ten virgins, this word is used. They took their lamps and went out to "*apantesin,*" that is, to meet and escort their bridegroom to return with him for the wedding feast.

The most common understanding of the rapture is that Christians will not face the tribulations coming to the earth. They will experience being caught up with the Lord in the air and will go to heaven as earth meet its destruction. However, a careful reading of all of the New Testament Scriptures clearly indicates that we will meet the Lord in the air, and soon after, we escort Him back to the earth. This is the second coming of the Lord to the earth with His bride, to set up His the kingdom of God on the earth. At that time, the kingdoms (or nations) of the world, will be the kingdoms of the Lord and of the Messiah, Lord Yeshua.

"*And then the sign of the Son of Man will appear in the sky, and then all the tribes of the earth will mourn, and they will see the Son of Man coming on the clouds of the sky with power and great glory.* ***And He will send forth His angels with a great trumpet and they will gather together His elect from the four winds, from one end of the sky to the other"*** (Matt. 24:30-31).

The book of Revelation says that heaven will come down to earth forever. It is my Scriptural belief that when the "*rapture*" is completed, the Lord will return with His Bride to the earth, as the

King of kings and the Lord of Lords. The radiant bride comes back with Him, and she will earnestly begin to disciple the nations in the kingdom and culture of heaven.

"Now it will come about that in the last days the mountain of the house of the Lord will be established as the chief of the mountains, and will be raised above the hills. And all the nations will stream to it, and many peoples will come and say, come, let us go up to the mountain of the Lord. To the house of the God of Jacob; that He may teach us concerning His ways and that we may walk in His paths. For the law will go forth from Zion and the word of the Lord from Jerusalem. And He will judge between the nations, and will render decisions for many peoples. And they will hammer their swords into plowshares and their spears into pruning hooks. Nation will not lift up sword against nation, and never again will they learn war" (Isa. 2:1-4).

"Then I saw a new heaven and a new earth; for the first heaven and the first earth passed away, and there is no longer any sea. And I saw the holy city, new Jerusalem, coming down out of heaven from God, made ready as a bride adorned for her husband" (Rev. 21:1-2).

It is time for the church to ask the Holy Spirit to teach her how to disciple the nations in the government of the Lord Jesus Christ, and as citizens of the kingdom of God on earth.

Intercession

We need to intercede for the church to get ready to disciple the nations. In the gospel of Matthew, soon after the Lord Jesus' apocalyptic prophecy of His second coming (see Matt.24), there is a parable about the faithful servant of the master. The context of this parable is the preparedness of the church when He returns to the earth.

"For this reason you also must be ready; for the Son of Man is coming at an hour when you do not think He will. Who then is the

faithful and sensible slave whom his master put in charge of his household to give them their food at the proper time? Blessed is that slave whom his master finds so doing when he comes. Truly I say to you that he will put him in charge of all his possessions. But if that evil slave says in his heart, 'My master is not coming for a long time,' and begins to beat his fellow slaves and eat and drink with drunkards; the master of that slave will come on a day when he does not expect him and at an hour which he does not know, and will cut him in pieces and assign him a place with the hypocrites; in that place there will be weeping and gnashing of teeth" (Matt.24:44-51).

This scripture is the command of the Lord to His church. The Lord will judge the church if she fails to comply with this. Peter gives importance to this command in this scripture:

"For the time has come for judgment to begin with the household of God. And if it begins with us, what will be the outcome for those who refuse to obey the gospel of God?" (1Pet.4:17).

The Third Stage: The Black Horse Rider

"*When He broke the third seal, I heard the third living creature saying, "Come." I looked, and behold, a black horse; and he who sat on it had a pair of scales in his hand. And I heard something like a voice in the center of the four living creatures saying, "A quart of wheat for a denarius, and three quarts of barley for a denarius; and do not damage the oil and the wine"* (Rev. 6:5-6).

Two symbols indicate the assignment that the black horse rider is given, as he moves through the nations:

1. He has a balance in his hand
2. A voice from the throne room calls out the rider's assignment: "*A quart of wheat for a denarius, and three quarts of barley for a denarius; and do not damage the oil and the wine.*"

The black horse rider is not the opposite of the white horse rider. God is stripping the nations and this is one more angelic being, coming from the presence of God in the throne room. This angel's assignment is to create famine on the earth. In a number of scripture passages in the Old Testament, we find God using famine to draw nations to His call and purposes. We could also see this as God's judgment upon the nations, and once again, He is taking away things that man needs most for sustenance of life on the earth, food and shelter.

It is not just a famine of two food commodities, wheat and barley, while there is a glut of oil and wine. It is symbolic of the entire food chain. There will be scarcities and shortages and the price of food will go up abnormally high. A normal person's monthly wage will not be able to get enough food on the table. A person's wage will supply very little for the niceties of life, like entertainment. This is represented by the "oil and wine" in the quoted scriptures. The black horseman brings with him a warning that more severe judgment is on the way. When God fails to get a response, after the red horseman visits a nation, He sends the black horseman. The balance, or scales in the hands of the black horse rider, is symbolic of the angel going through the nations "weighing the response" to the Word of God.

How has the word of God been received in this place?

How has the nation responded?

How has the church responded?

Revelation 3 reveals the black horse rider as he moved through the seven churches of John's time. The Lord rebuked those churches for their apostasy and lack of faith. Scripture records this, so that every successive age can see the pattern and respond to God in a manner that pleases Him. We are now in the last church age, and will now see the ultimate fulfillment of the word of God.

The black horse rider comes to weigh in the balance the response of our nation, cities and churches to the gospel.

Has the Black Horseman Visited the Nations?

Sometimes the record of events on the earth, give us a picture that could help with our discernment.

Dec. 10, 2010: Part of New Zealand's North Island Declared Drought Area

Growth in New Zealand's milk output had slipped below normal, for the season up to the end of November, hampered by a cold spring.

Dec 10, 2010: Middle East Wheat Faces Drought Threat

Middle East wheat production next year is threatened by a drought during the main planting period that is weighing on crop development, a major forecaster said. Wheat-growing regions in southern Turkey, Syria, Iraq and northern Iran received less than 20 percent of normal rainfall in the 60 days.

Dec 9, 2010: Crops worth millions washed off in Chhattisgarh (India) rains

After braving a poor monsoon, farmers in Chhattisgarh are again facing testing times as unabated rains since Tuesday have destroyed their paddy crops stored in open grounds.

Dec 9, 2010: GRAIN and fruit growers are on the brink of disaster across southeastern Australia.

Eastern and Southeastern Australia has had incessant rain and cyclonic storms that has devastated the economy. Damage is estimated at about $2.1 billion, which has been wiped from grain growers' income in NSW and Queensland.

Dec 9, 2010: 'Stealth bomber' fungus sneaks into crops: study done in Europe

20 to 40% of the world's harvest is lost, because of pests like the "stealth bomber," due to its adaptability and difficulty in its detection.

Growing Threat to Mexico's Crops – Multi-Billion-Dollar Consequences for US

Crops Ruined, 6 Nashik (India) Farmers Commit Suicide 2010 'one of worst' years for disasters (From: *AFP* January 24, 2011)

- Cost of natural disasters $109 billion. 2010 U.N. report
- China drought threatens water supplies: state media, Jan 24, 2011
- Food prices to rise sharply as more struggle to put meals on their tables
- Are We at the Beginning of a Global Food Crisis? World's Food Chain Maxed Out Extreme Winter. 10,000 Cows Drop Dead in Vietnam
- Drought And Extreme Cold Throughout China

The other outcome of the black horse rider's visit to the nations is the economic effects on the nations. Shortages in the food chain lead to increased prices, and that leads to unemployment. Food riots recently seen in Bangladesh were the result of the combination of food shortages, high prices and unemployment. The current global unemployment problem is so huge that the total number of jobless in the ten most populous nations in the world totals 1.1 billion. That is only slightly smaller than the population of China (1.3 billion).[28] Most secular analysts predict riots in nations in the years ahead.

Intercession

Some theologians think that the world is now seeing the visitation of the black horse rider. Like Abraham, the church, in this time of partial judgment of nations, must cry out for the mercy of God on the poor and marginalized. How can the coming of the pale horseman, the last of the four horsemen, the rider of mass deaths, be delayed?

We will need to see much more intercession go up to God from our cities and nations. The Lord needs more churches praying in unity of purpose. This will bring a harvest of souls, released by the

[28] See web site for details: http://www.huffingtonpost.com/2010/06/23/the-15-nations-with-the-h_n_622174.html

visitation of the black horseman. The basic need for food and work will lead people to seek God's mercy. Jesus said the sunlight falls on the good and the evil people alike. People of all walks of life, will experience hardship that results from the visitation of the black horse rider. Yet God's covenant promise is that He will care for those who love Him. We can ask for the Father's mercy our nation, and pray that she turns from her wicked ways, before the visitation of the black horseman.

"***At one moment I might speak concerning a nation or concerning a kingdom to uproot, to pull down, or to destroy it; if that nation against which I have spoken turns from its evil, I will relent concerning the calamity I planned to bring on it****. Or, at another moment I might speak concerning a nation or concerning a kingdom to build up or to plant it; if it does evil in My sight by not obeying My voice, then I will think better of the good with which I had promised to bless it. So now then, speak to the men of Judah and against the inhabitants of Jerusalem saying,* ***thus says the Lord, Behold, I am fashioning calamity against you and devising a plan against you. Oh turn back, each of you from his evil way, and reform your ways and your deeds***" (Jer. 18:7-11).

This is God's covenantal promise to stay the destroying hand of the rider of the black horse, if people will repent of their evil deeds. However, history records that man very rarely repents of his evil ways, even after many warnings.

"*But they will say, 'It's hopeless! For we are going to follow our own plans, and each of us will act according to the stubbornness of his evil heart'*" (Jer. 18:12).

"*For My people have forgotten Me, they burn incense to worthless gods and they have stumbled from their ways, from the ancient paths, to walk in bypaths, not on a highway, to make their land a desolation, an object of perpetual hissing; everyone who passes by it will be astonished and shake his head. Like an east wind I will scatter them before the enemy; I will show them My back and not My face In the day of their calamity"* (Jer. 18:15-17).

The Fourth stage: The Pale Horse Rider.

"When the Lamb broke the fourth seal, I heard the voice of the fourth living creature saying, 'Come.' I looked, and behold, an ashen horse; and he who sat on it had the name Death; and Hades was following with him. Authority was given to them over a fourth of the earth, to kill with sword and with famine and with pestilence and by the wild beasts of the earth" (Rev.6:7-8).

The fourth horseman, who rides the pale horse, has been given the following command

1.) Authority to kill with the sword, this is a strong indication of wars, riots, and civil unrest in the nations
2.) Deaths due to famine, pestilence, and natural disasters, including from wild beasts.
3.) The rider of the pale horse has a name: "Death." In 1 Chron. 21:16, the angel of the Lord is described as standing between earth and heaven with a drawn sword in his hand: *"David looked up and saw the angel of the Lord standing between heaven and earth, with a drawn sword in his hand extended over Jerusalem. Then David and the elders, clothed in sack-cloth, fell facedown." The Angel of the Lord slew 185,000 Assyrians"* (also a reference is made in Isa. 37:36).

    God's heart changed towards Jerusalem when He saw the destruction of the people: "*When the angel stretched out his hand toward Jerusalem to destroy it, the Lord relented from the calamity and said to the angel who destroyed the people, 'It is enough! Now relax your hand!' and the angel of the Lord was by the threshing floor of Araunah the Jebusite*" (2 Samuel 24:16).
4.) Hades, the place where souls that have rejected God and await the final judgment. It follows the angel of Death.

5.) The rider of the pale horse has authority over the fourth part of the earth. This indicates a "partial" judgment of the earth and not the final judgment. God is still merciful more than we can imagine. However, man cannot twist the hand of God because He is merciful. His day of reckoning has come, so also the judgment of the nations.

Jesus said in Luke 17:26-27, "*And just as it happened in the days of Noah, so it will be also in the days of the Son of Man: they were eating, they were drinking, they were marrying, they were being given in marriage, until the day that Noah entered the ark, and the flood came and destroyed them all.*" In other words, Jesus was emphasizing that in Noah's days, it was "business as usual," or, people were "indifferent" right up until the floods came. However, the time has come, when business will not be "as usual" anymore, for the day of reckoning has come, by the visitation of the pale horse rider to the nations.

If the world continues to be indifferent to sin, and repentance and return to the Lord is not forthcoming, the fourth pale horseman will be sent by God. There will be many deaths due to a number of factors: Wars, famine, and natural disasters. This may annihilate a quarter of the world's population.[29]

It has been amazing to read about the recent unrest and riots in Islamic dominated nations. Some of you may recall that during the early months of 2011, a video captured the image of a pale green horse and rider, who was riding over the riotous, violent streets of Cairo, Egypt. Millions watching on their computers and on "YouTube" saw this supernatural visitor, as death and violence reigned on the streets of Egypt. When Jill and I (separately) saw this scene, we were shocked. It happens that the Greek word that John used in Revelation 5, for "pale horse," was actually the word *"chloros,"* which means a shade of yellowish-green. It is a sickly kind of green, like death. From this word, we derive the English word, "chlorine" and "chloroform." Most people do not know that

[29] Take for example the earthquake and tsunami in Japan on Marsh12, 2011. Apart from significant deaths and missing people, the nations is now under a nuclear threat from damaged reactors.

John saw a pale GREEN horse, and therefore, no one could have falsified this video footage in this way. They would have chosen a pale (light cream, or light grey ash) color, but certainly not green. This is perfect evidence that the four horsemen are indeed riding the nations of the earth in this last generation. The Lord allowed us to see it on camera. We need to pray for the Islamic nations, that the Lord of the harvest will draw forth a harvest for Himself from these nations, during the visitation of the fourth horsemen.

"*For nation will rise against nation, and kingdom against kingdom, and in various places there will be famines and earthquakes. But all these things are merely the beginning of birth pangs*". (Matt. 24:7-8).

Here is some evidence that the pale horse rides across the nations of the earth.

Deaths by wars, massacres, slaughters and natural disasters.

- **From the end of the WWII in 1945 to 2008, the world has seen 313 conflicts, resulting in 92 million people killed.** Estimated totals are for the entire 20th Century.[30]

  More people have been killed by conflict in the past 60 years than double the number of victims killed in World Wars I and II combined.

Deaths due to religious persecution:

- **David Barrett**, ***World Christian Encyclopedia*** **(2001):** This book is the standard reference work for religious statistics of all kinds, and both Britannica and the World Almanac cite from it. **Christian martyrs in the 20th Century only: 45.5 million.**[31]

[30] For full details of text: http://www.insidejustice.com/law/index.php/intl/2010/03/31/p256.

[31] Excellent web site for world statistics: http://users.erols.com/mwhite28/warstat8.htm

Deaths due to Communism

- **Stephane Courtois**, in his ***The Black Book of Communism:***[32] Victims of Communism estimated 85-100 million.

Viruses and plagues in the 20th century:

- Michael Oldstone, is a Member (Professor) at the Scripps Research Institute, where he directs a laboratory of viral immuno-biology. He is currently a member of the World Health Organization steering committee concerned with the eradication of measles and poliovirus, an editor of the journal **Virology**, and the recipient of numerous scientific honors. He was also Scientific Counselor for the intramural program of the Allergy and Infectious Disease Unit of the National Institute of Health and was recently elected to the Institute of Medicine of the National Academy of Sciences. In his *book "Viruses, Plagues, and History,"*[33] Mr. Oldstone focuses on a few of the most famous viruses humanity has battled, beginning with some we have effectively defeated, such as smallpox, polio, and measles. **Nearly 300 million people were killed by smallpox in this century alone — more than were killed in all the wars of the twentieth century combined.**

AIDS:

1. As of 2009, deaths due to the AIDS virus has reached over 25 million over the last 28 years since its outbreak in 1981.
2. There are an estimated 33.3 million people now living with HIV.[34]

Asian and Hong Kong flu pandemics:

- The Hong Kong flu became the third flu pandemic of the 20th century. The World Health Organization estimated that a total of 1.5 million died in the Asian and Hong Kong flu pandemics.

---

[32] Havard University Press, 1999, also from http://users.erols.com/mwhite28/warstat8.htm

[33] Oxford University Press, 1998. First paperback edition, 2000

[34] Worldwide HIV and AIDS statistics: http://www.avert.org/worldstats.htm

Deaths due to natural disasters: A total of 3.5 million people were killed in the 20th century disasters such as floods, earthquakes, and volcanoes (not included are deaths due to drought or famine).

Deaths in recent years:

2003 **Europe Heat Waves**- 35,000 dead; **Iran Earthquake**- 31,000 to 43,000 dead.

2004 **Asian Tsunami**- 225,000 to 275,000 dead.

2005 **Hurricane Katrina**- 1,800 dead; **Pakistan Earthquake**- 75,000 dead.

2008 **Burma (Myanmar) Cyclone-** 146,000 dead; **China Earthquake-** 70,000 dead. **Guatemala/El Salvador Hurricane-** 1,638 dead.

2009 **Global Swine Flu**- 11,800 dead.

2010 **Haiti Earthquake**- 230,000 dead; **Chile Earthquake**- 750 dead; **Pakistan Flooding**- 1,800 dead; **Russia Heatwave**-15,000 dead; **European Cold Wave**; over 30 dead.

2011: significant deaths have occurred in the Christchurch, New Zealand earthquake, and the tsunami in Japan.

Carl Haub [35] is a senior demographer at the Population Reference Bureau in Washington, D.C. He is the author of the 2010 World Population Data Sheet, which showed that during the 20th century, there were approximately 9.8 billion births. Add to that another 1.66 billion alive at the beginning of the 20th century. The total population, which should have been alive at the end of that century, was approximately 11.46 billion. However, by the end of the 20th century there were actually only 6 billion people alive. In other words, 5.46 billion people or close to 50%, died. This figure includes those who died naturally, of old age, but old age alone would not account for so many deaths.

The huge number of deaths in the 20th century alone causes us to ponder if man has the wisdom to "subdue the earth," as God had entrusted him. We can acknowledge that man's increase in knowledge has achieved great benefits for the human race. Yet the insanity

[35] Read the article in http://www.prb.org/Articles/2002/HowManyPeopleHaveEverLivedonEarth.aspx

to kill one another, to war, has not left human desire. Will humanity learn anything from the pale horseman's visit, which results in annihilation of a fourth of the world's population? Can we not see the glaring statistics and read the writing on the wall? Jill had a dream, in which she saw the back of God, standing at the rock wall of a huge cavern. He was writing on the wall. This was prophetic.

In the book of Revelation, John wrote about the four horsemen twenty centuries ago. He warned that God would bring limited judgments as a warning of the impending final judgments. Man is not in control of the destiny of the earth. God is, and He does so with a purpose. Through worship, warfare, and intercession in heaven, God controls the destiny of the earth, in order to fulfill His purposes. Since God is in control, should the prophet remain silent? **The Lion has roared; who can but prophesy?** (see Amos 3:8). Does the intercessor stop praying? Obviously not! For those who love the return of the Lord Jesus Christ to earth, they must join heaven in the work of restoration and renewal of the earth. The prophet must continue to stand before the throne of God and hear the Lion roar over the nations of the earth, which initiates their judgment. The prophet must also hear the Lamb speak to the church, His bride, and turn her focus to His soon return as King of kings and Lord of lords. We must understand what God is doing, and join Him in through our intercession. God is shaking the earth as He said He would (see Hebrews 12:26-29). We should pray through His judgments, crying out that He will render mercy on those who diligently seek His face. *"In wrath, remember mercy!"* (Hab. 3:2).

Sometimes it is a hard decision to pray for the Lord's return. Effectively, this will mean praying with heaven for the judgment of the nations. God's judgment of the nations sees destruction and death of the effects of sin, yet mercy and compassion on those who repent and seek Him, restoration and renewal of the earth, His final redemptive act.

The prophet Elijah came against Israel's apostasy, as she vacillated in her faith between two sides, Yahweh and fallen man:

"Elijah came near to all the people and said, ***"How long will you hesitate between two opinions? If the Lord is God, follow Him; but***

if Baal, follow him." *But the people did not answer him a word"* (1Kings 18:21).

This is the time for the church to make her decision not scuffle between the world and her Lord, trying to get the best of both worlds. We must petition the Lion of the tribe of Judah, to tear down the modern day altars of "Baal," which were set up in their nations. In this time and age, how are we to define what the altars of "Baal" are in our nation? Consider this scripture:

"We are destroying speculations and every lofty thing raised up against the knowledge of God, and we are taking every thought captive to the obedience of Christ" (2Cor. 10:5).

In everything that "exalts itself against the knowledge of God," there an altar to Baal that has been erected. These altars need to be pulled down through worship of our God and through warfare and intercession. You will find there is hardly anything that a nation does that exalts the name of King Jesus! Most of our cities are a multi-religious mix; hence, the Lord's name, or anything to do with the one true God, the Lord Jesus, governments relentlessly cast out in order to maintain a "political correctness" stance. Even if this leads to the judgments of God on our cities, of the magnitude we have seen in recent times in New Zealand and Japan.

In many nations, many of the institutions do not know the difference between culture and religion. My wife and I were at a public library, and we asked the librarian why an altar was raised to "Kali," one of the pantheistic Hindu goddesses, in the library. The librarian said that it was a celebration of a cultural festival season, and that they encourage ethnic festivity. My wife told her that the display of the huge idol and altar of incense was not a cultural message being sent out by the library, but a display of their ignorance between culture and religion. My wife then asked her if she would allow a Christmas display of the birth of the Lord Jesus Christ in the library during the season of Christmas, and which is a public holiday in New Zealand. The librarian gave an emphatic NO, saying, "That's religion, and we don't preach any religion in a public government funded library. We put up Santa Claus and a Christmas tree decorated with lights in the library to celebrate Christmas." Anything to do with Christianity or the Lord Jesus, is designated as religion, but "Kali", a part of the Hindu pantheon of gods and goddesses is culture! This is an example of the abject display of ignorance in New Zealand of what comprises religion and what culture is all about. In such circumstances of blatant stamping out of the name of the Lord Jesus by public funded government run institutions, attracts a visitation of the pale horse rider to our cities in New Zealand, seems to me, in place. What happened in Christchurch proves a point in this regard.

Western society all over the world, has raised altars to "Baal" in every nook and corner of public life. Another instance of how altars to "Baal" are raised in nations, took place when I was surprised to find an idol in a commercial bank. As I entered the commercial bank, in Auckland,, I was immediately confronted by a huge statue of another of the Hindu pantheon of gods, the "elephant-god," named "Ganesh." I asked the bank manager why, in a country like New Zealand, where Christian religious sentiments were not allowed to be expressed in commercial establishments, was such an open display of non-Christian religion openly expressed?

Instantly, he answered, "The elephant-god is the god of wealth. That's what all of us want, wealth. That is what a bank is all about, the creation of wealth. This is a symbol of the bank's core functionality, and it has nothing to do with religion. We are celebrating the giver of wealth, which is synonymous with the business of banking, creating wealth. I can't see anything wrong with building excellent banking relationships, through the symbols of wealth creation." Wow, I thought, what a way to introduce the Babylonian system of perverted thinking back into modern day society.

I was surprised at the answer, since it came from a man who professed to be a practicing Christian, by that I mean, a church member. He said it was a "policy decision." He was not in a position to decide what was right or wrong. Most policy decision-makers, chasing wealth, power and fame, are the ones who raise the altars to Baal. Unfortunately, like in this example, their employees are made to bow down to these altars, and to it to hold on to their jobs in an economy going down-hill. I reminded him of the time that Israel built the golden calf and danced around it. They worshipped mammon, the god of wealth and money, and all the evil that goes with it. I said, "Times have not changed since then, mate." He smiled at me, shook my hand and ushered me out of the bank as my discussion was not welcomed in the bank.

New Zealand prides itself on its "western culture," which considers itself modern and "forward thinking." Probably, this is the plight of all of western civilization all over the world. From Wall Street, to the bank in New Zealand, men have raised altars to "Baal." Who will tear these altars down? If the church, which has this mandate, power and authority from the Lord, does not tear them down, then she, like Israel, must repent for not doing what she ought to do.

"*Elijah came near to all the people and said, 'How long will you hesitate between two opinions? If the Lord is God, follow Him; but if Baal, follow him.' But the people did not answer him a word"* (1Kings 18:21).

Intercession

God is very patient, and while He is yet patient, we need through worship, warfare and intercession open the spiritual eyes of the blind, the spiritual hearing of the deaf, the spiritual lame will walk again. We pray that rivers of salvation will flow out to the sick heart of humanity. We do not just pray for physical healing, but for the healing of the sick soul of man, who is in rebellion against God. We pray that the nations will repent of their evil, governments will open their eyes, and that brave leadership will emerge to do what is right and pleasing in the eyes of God.

Each passing year will see more of God's judgment on the nations. We pray for the people, who experience God's judgment on the nations that will bring hardship and stress. In these hard times, their hearts will turn to God. Our Heavenly Father still stands at the gates, as He stood looking out for His prodigal son. He is hoping that with each passing day, as darkness covers the earth, more people will find their way into the mansion He has prepared for them, through the redeeming blood of His Son. **God's hand is not too short that it cannot save in the midst of the judgment of the nations.** Indeed, nations can save themselves from the wrath of God if, like Nineveh, they would repent and return to the Lord. In the absence of a nation's response to follow the Lord, by its godless laws and way of life, they have chosen to invite the four horsemen. Three of these bring God's judgments on the nations. The psalmist in Psalm 2 gives us their response to the love of God:

"*Why are the nations in an uproar And the peoples devising a vain thing? The kings of the earth take their stand and the rulers take counsel together against the Lord and against His Anointed, saying, 'Let us tear their fetters apart And cast away their cords from us!' He who sits in the heavens laughs, the Lord scoffs at them. Then He will speak to them in His anger and terrify them in His fury, saying, 'But as for Me, I have installed My King upon Zion, My holy mountain.'*

"I will surely tell of the decree of the Lord: He said to Me, 'You are My Son, today I have begotten You. Ask of Me, and I will surely give the nations as Your inheritance, and the very ends of the earth as Your possession. You shall break them with a rod of iron, You shall shatter them like earthenware.

"Now therefore, O kings, show discernment; take warning, O judges of the earth. Worship the Lord with reverence and rejoice with trembling. Do homage to the Son, that He not become angry, and you perish in the way, for His wrath may soon be kindled. How blessed are all who take refuge in Him!'" (Psalm 2:1-12).

The church must not give up hope in prayer, for even at the very last hour, peoples and nations will repent and return to the Lord.

News headlines below reveal that the four horsemen of the Apocalypse are moving through the nations, looking for the response of man to God. More changes in weather patterns that will continue to exert pressure on the economies of nations. The current cold wave and winter blizzards, experienced in many nations in the northern hemisphere, indicate this.

"Yardsticks of Snow for California:" A minimum of 5 feet of snow through Sunday.

Britain in for Worst Blizzard in 100 Years: Drifts of up to 6 feet expected, as 60mph Arctic gales batter the country. The Polar freeze is set to cost their economy £13.2 billion by Christmas, wiping out 1% of the UK's annual turnover, said loss experts.

"There's a mini ice age coming," says man who beats weather experts, December *21, 2010.* Piers Corbyn not only predicted the current weather, but he believes things are going to get much worse.

Germany brought to near standstill by 12 hours of solid snowfall: On the A9 autobahn leading into Munich, the traffic jam into the city reached 18 miles long. Several people were killed in the accidents and dozens hospitalized.

"Remember my affliction and my wandering, the wormwood and bitterness. Surely, my soul remembers, and is bowed down within me. This I recall to my mind, therefore, I have hope. The Lord's loving kindnesses indeed never cease, for His compassions never fail. They are new every morning; great is Your faithfulness" (Lam. 3:19-23).

We can praise and thank God that even in His judgments, His mercies and compassions are still new every morning, and He will provide for those who love Him. Surely, goodness and mercy will follow after those who will seek the Lord and His salvation, even in the proverbial eleventh hour. Praying Psalm 23 for nations is one way to stand on God's word, especially for His saving hand and loving kindness, in the time of judgment. We direct our intercession to God who is a loving Father, who is merciful even in His judgments. We can also pray that through God's righteous judgments, people will return to God and find in Him a safe shelter. *"When your judgments come upon the earth, the people of the world learn righteousness"* (Isa. 26:9b, NIV).

In Summary

We must remember that God is moving in accordance with His revealed word.

"*For My thoughts are not your thoughts, nor are your ways My ways, declares the Lord... So will My word be which goes forth from My mouth; it will not return to Me empty, without accomplishing what I desire, and without succeeding in the matter for which I sent it*" (Isa. 55:8, 11).

How the Lord will restore the earth to its pristine position, man cannot imagine or dictate. Sinful man has messed up the earth that God created. It will need God alone to restore it. One thing we know: When God restores and renews the earth; it will truly be a marvel in our eyes. The curse on the earth will be removed, so also the thorn and the thistle. It shall not be a hard and difficult environment. Sin, death and decay will be removed, as well as their effects on creation. Neither any government, nor any man or human institution will be able take credit for it. The restoration and renewal of the earth is the gift of God for those who love Him and seek diligently to please Him.

In the Father's love for His eternal Son, Our Lord Jesus, and His bride, God has given a sure word for this age.

"*...but just as it is written, 'things which eye has not seen and ear has not heard, and which have not entered the heart of man, all that God has prepared for those who love Him.'*" (1Cor. 2:9).

"*Then I saw a new heaven and a new earth; for the first heaven and the first earth passed away, and* ***there is no longer any sea.***" (Rev. 21:1).

"For you will go out with joy and be led forth with peace; the mountains and the hills will break forth into shouts of joy before you, and all the trees of the field will clap their hands. Instead of the thorn bush the cypress will come up, and instead of the nettle the myrtle

will come up, and it will be a memorial to the Lord for an everlasting sign which will not be cut off." (Isa. 55:12-13).

"*...and He will wipe away every tear from their eyes; and there will no longer be any death; there will no longer be any mourning, or crying, or pain; the first things have passed away." and He who sits on the throne said, 'Behold, I am making all things new.' And He said, 'Write, for these words are faithful and true.'*" (Rev. 21: 4-5).

We need the Father's heart in our intercession. His concern is our concern, and His plans and purposes must be our plans and purposes. This is a very important principal in intercession. If we do not have this perspective, the coming judgment on the nations may cause us to stray from the purposes of God. We may even find ourselves praying against His purposes.

Here are a few guidelines. The Holy Spirit will be your best teacher.

Current events on the earth indicate that God is shaking everything that glorifies fallen man. It is a very painful process, as God destroys landmarks that speak of man's glory. It is also a costly process in terms of human loss and other physical and financial losses. We pray that those who diligently seek Him will experience His mercy. We do not pray that the shaking of the nations will stop, because we cannot pray against the word of God. The word of God has set these processes into motion. Remember the rebuke that the Lord Jesus gave Peter. He rebuked him for coming against God's will:

"*From that time Jesus began to show His disciples that He must go to Jerusalem, and suffer many things from the elders and chief priests and scribes, and be killed, and be raised up on the third day. Peter took Him aside and began to rebuke Him, saying, 'God forbid it, Lord! This shall never happen to You' but He turned and said to Peter, 'get behind Me, satan! You are a stumbling block to Me; for you are not setting your mind on God's interests, but man's'*" (Matt. 16:21-23).

We are now witnessing the ultimate fulfillment of all of the Lord Jesus' end-time prophecies in Matthew 24, Luke 21 and Mark 13. The ancient prophets prophesied the tearing and building up of the nations. While the church on earth prays for the fulfillment of God's declared word, she must also pray for all those who will experience suffering, due to these end time events. God's heart is for saving humanity from the devastating effects of His judgment of the nations. We must intercede for the people affected by these earth-stripping events, and pray for a conversion of heart and a return to God.

In the beginning of 2010, I was praying with a group of Christians in a city in New Zealand. They conducted a 24/7 intercessory worship prayer vigil. The Lord gave me a vision of the Father, standing at the gate, looking at a cloud of dust on the far horizon, waiting for His children to come home. I could see the pleasure in the Father's eyes as He gazed with hope at that cloud of dust. It meant that there was a crowd of people heading towards the gate. He said loud enough for me to hear: *"They must come soon, for darkness will soon fall, and they may lose the way, and the gates will be closed."* The vision suddenly changed and I saw myself in a farm with a spade in my hand, digging the ground that was dry, arid and hard. The city (where I was praying) was a rich farming community. I immediately understood the vision. It was a call to pray for that city. The vision seemed to indicate that the hearts of the people were dry, arid and hardened by sin, as well as rejecting of God. I wondered were these the people who had refused to come into the Father's house, which Jesus has prepared for them. I wept in intercession for that city. I remembered that even Jesus wept over Jerusalem. I prayed that the city would be spared the fate that Jerusalem experienced, for it did not acknowledge the time of the Lord's "visitation" (see Luke 19:44).

As the four horsemen move over the earth today, we are seeing such fearsome climatic changes, which we have covered. The Father is still waiting at the gate, soon it will be nightfall and He will command to shut the gates. A new dawn of righteousness, peace and joy will not arise on the earth until the King of kings and the Lord of lords, the Lord Jesus Christ, returns with healing for the nations in His wings.

A Time for National Repentance

National repentance is about a nation recognizing that it has strayed away from God's law and rejection of God's love in Christ Jesus our Lord, and returning to follow the Lord. The government and the people must desire to put God first. The lack of national repentance often brings human suffering, damage, brokenness and loss in its wake, through natural calamities. God can change the destiny of a people or a nation. Repentance by a nation moves the heart of God. The church, like Ezra and Nehemiah, needs to call the nation to national repentance. **The church is the prophetic voice in the nation.** She cannot and must not stop calling the national leaders and the entire nation to repentance, when they have gone far from God's laws and ways. The voice of the church today is weak and low in intensity, when it comes to calling the nation to repentance. A divided church has neither face nor voice in a nation anymore. Exploitation of the nation continues, unabated, shamelessly.

How long will God allow a godless society to continue its march of exploitation of the earth, and of the less fortunate of society? In the destruction of native forests, and in the senseless killings by drug cartels, the poor and the marginalized suffer. The cunning of banks and insurance companies rob the ordinary workmen of more than 50% of their wages, through the unjust system of mortgage payments. Over an average worker's lifetime, he will pay 500 times over the cost of his house, by way of bank interest; this is a glaring injustice. In a recent article in the New York Times, December 22, 2010, the news article heading read: "Banks accused of illegally looting homes." When the church does not take up the cry of her people, God will come to their aid. The worldwide financial crash of 2008 began with the mortgage market.

Governments are to serve those who elected them, rather than lord it over them. The cry of those who suffer this injustice has reached the ears of God. He will come to deliver them. He is the God of righteousness and justice. Man in his greed has not learned his lessons; neither have the governments of the nations. Their greed and pride are an abomination to God. If the church on earth does not call their governments to repentance, God responds to the intercession in heaven, and He takes back what belongs to Him, the entire earth.

He has initiated the "end time protocol." The four horsemen are on the march and no army, no "one world government," no technology, and no force on the face of the earth can stop this. Why are nations willing to be down trodden by the horsemen of heaven, as they move to bring God's righteous but partial judgments on the earth? Is it too hard to kneel before the cross of Christ and repent of all unrighteousness? The Word of God tells us why:

"*For the wrath of God is revealed from heaven against all ungodliness and unrighteousness of men who suppress the truth in unrighteousness, because that which is known about God is evident within them; for God made it evident to them. For since the creation of the world His invisible attributes, His eternal power and divine nature, have been clearly seen, being understood through what has been made, so that they are without excuse. For even though they knew God, they did not honor Him as God or give thanks, but they became futile in their speculations, and their foolish heart was darkened. Professing to be wise, they became fools*" (Rom.1:18-22).

"*For the word of the cross is foolishness to those who are perishing, but to us who are being saved it is the power of God. For it is written, 'I will destroy the wisdom of the wise, and the cleverness of the clever i will set aside.' Where is the wise man? Where is the scribe? Where is the debater of this age? Has not God made foolish the wisdom of the world? For since in the wisdom of God the world through its wisdom did not come to know God, God was well-pleased through the foolishness of the message preached to save those who believe. For indeed Jews ask for signs and Greeks search for wisdom; but we preach Christ crucified, to Jews a stumbling block and to Gentiles foolishness, but to those who are the called, both Jews and Greeks, Christ the power of God and the wisdom of*

God. Because the foolishness of God is wiser than men, and the weakness of God is stronger than men" (1Cor. 1:18-25).

This is the Hour to Pray for Your Government

The world is moving into a time when governments will find it hard to rule their nations. They will need wisdom to handle these difficult days. The church has a role to play in these times. Like Joseph and Daniel, she will have a role to play in the ways of governing. Governments will seek after Christians to guide them in secret. Godless leaders will even detain or arrest Christians, in order to use them for wisdom and guidance from the Lord, similar to the cases of Pharaoh and Joseph, Nebuchadnezzar and Daniel.

"Then the kings of the earth and the great men and the commanders and the rich and the strong and every slave and free man hid themselves in the caves and among the rocks of the mountains; and they said to the mountains and to the rocks, 'Fall on us and hide us from the presence of Him who sits on the throne, and from the wrath of the Lamb; for the great day of their wrath has come, and who is able to stand?'" (Rev. 6:15-17).

Will the church answer the call? Events are unfolding faster than one can imagine. Wisdom to govern and rule by God's standards will be very hard to find. Watch how chaos will be the rule of the day! **Our God is the Lord of chaos.** From out of deep darkness of chaos of unspeakable proportion, He will draw out the new heaven and the new earth for Himself, and will create a dwelling place for those who love Him. The increase of His government will have no end! (See Isa. 9:6).

Watch the kind of world scenarios one may expect to see unfold in their final biblical patterns.

- Before our very eyes, riots in Middle-Eastern nations are overthrowing regimes that reigned for decades in peace and harmony. The rider of the red horse is taking away that peace. What will happen to these nations that cry for democracy?

It is hard to perceive fundamental Islamic nations having a democratic government. It is not in their mental and cultural make-up, nor is it within their spiritual mandate, to be democratic. The more correct word is domination. The one group that can dominate all other groups reigns in power. There is only one thought that unites them all: their hatred for Israel. In these events, we see biblical prophecy unfolding. That is the ultimate unity of Islamic nations, preparing to carry out their one goal, the destruction of Israel.

- We shall see the weakening of the western nations financially and politically. National problems weaken the nation from within, as the Babylonian system crashes, giving opportunity for Islamic fundamentalists to accelerate their programs of generating the psychosis of terror in their victims. Western nations will experience religious riots from fundamental Islamists, whose focus is to put stress on the relationship with Israel. As western nations withdraw their allegiances, using resources to fix their own internal problems, Israel is left without friend, with no covenant brothers. She has to fend for herself without allies. Iran will see this as an opportunity to flex its muscles. Israel has only one Person to whom she can turn. The Lord Yeshua, their brother, their Joseph, long lost and forgotten, will reveal Himself to them by his loving care and provision. He has placed His covenant cup in their food basket, to drink the covenant drink with them.

- Food shortages and high prices will drive every nation to the brink of anarchy. Climate will continue to play a major role in food shortages.

- New world-wide alignment of nations, or grouping of nations, will begin to take place. The world is on the brink of a nuclear war, as it leads to the battle to control the spiritual capital of the world, Jerusalem.

One night I had a vivid dream. There was a man in the market place, like the town crier of old, and he shouted: "A loaf of bread for a denarius." This dream woke me up from my deep sleep. Later in the day, I looked up the bible and this passage related to my dream:

"*And I heard something like a voice in the center of the four living creatures saying, 'A quart of wheat for a denarius, and three quarts of barley for a denarius; and do not damage the oil and the wine'*" (Rev. 6:6).

On analyzing the prophetic scriptures, this is what I began to see in the message:

"A voice in the center of the four living creatures saying..." This voice coming from the center of the four living creatures, comes from the throne, which is surrounded by the four living creatures. Hence, God himself utters these words.

"A quart of wheat for a denarius, and three quarts of barley for a denarius." My bible's footnotes say that a denarius was like a day's wage, and it was not enough to buy a loaf of bread. Famine is coming and will affect many parts of the world.

"...do not damage the oil and the wine." Symbolically, the oil and the wine represent the anointing of the Holy Spirit in the church. This speaks of the coming glory in the church that the prophets predicted, which would be greater than its former glory. This shows us that the Lord will not only protect His people from famine, just as Yahweh protected Israel when she was in the wilderness, but the church will come into that time of doing the "even greater works" of God. (See John 14:12).

The true church must focus on the Lord in these times of difficulty, as her only source of sustenance. "*You shall charge the sons of Israel, that they bring you clear oil of beaten olives for the light, to make a lamp burn continually*" (Ex. 27:20). **Oil kept the lamps burning in the Holy Place; intercessory prayer is that oil needed to sustain the church in her role in the end times. She must be the house of prayer for all nations.**

We need God's grace to know how and what to pray, as we stand at the brink of what is to come. God has been so gracious to give us warning after warning. Do we hear His warnings in His word? Recently, there have been some unusual happenings in the world. We find that fish of one particular species, are dying by the millions, and birds (of one particular species), are falling from the skies, dead in their thousands. This has baffled scientists, who cannot find the answers. January 2011, saw this happen in different parts of the world:

January 27

Arkansas: Birds Died of Blunt Force Trauma

Grief of the Penguins: Scores of Penguins bowed in mourning after deaths of their chicks

January 26

Mass Death of Birds and Fish: Is There a Cover Up?

Birds Vanishing in the Philippines

January 24

Top USDA Bee Researcher Also Found Bayer Pesticide Harmful to Honeybees

Call to Ban Pesticides Linked to Bee Deaths Worldwide

Bird Death Is Spreading

55 Buffalo Die Mysteriously on NY Farm

January 20

Bees Facing a Poisoned Spring

Boat Harbor Canadians Baffled by Dead Seals, Fish

What Is Killing Portugal's Octopuses?

Thousands of Dead Octopuses Wash Up on Portugal Beach

200+ Dead Birds Found in South Dakota City

More Dead Birds Found, This Time in Scottsboro, Alabama

January 16

200 Dead Cows Found in Wisconsin Field

Over 9,000 Buffalo and Cows Freeze to Death in Vietnam

Mass Fish Kill in Iranian Sector of Caspian Sea; Blamed on Bio Weapons?

January 14

Tags Created to Help Penguins Could be Killing Them

January 12

Mystery Deepens: Radar Captures Mysterious Turbulence over Beebe, Ark. as Doomed Birds Take Flight; Dozens of Dead Birds Found in Missouri

January 11

Major, Bizarre Fish Die-Off along Chicago Lakefront

100+ Dead Birds Found Off California Highway

January 10

Native Bees Taking a Hit, Too

South Korea Buries One Million Pigs Alive

January 9

Swarms of Dead Birds Found in Tennessee

January 6

2 Million Dead Fish Line Chesapeake Bay

Bird and Animal Deaths around the world

Leaked: EPA Knowingly Approved Bee-Killing Pesticide

January 5

MUFON Links Recent Bird, Wildlife Deaths to Alien Technology

Faenza, Italy: The Rain of Death: Turtle Turns Blue

First Blackbirds in Arkansas, Now Jackdaws Drop in Sweden. A Common Cause?

Hundreds of snapper fish washed up dead on New Zealand's Coromandel Peninsula beaches

Dead Birds and Fish: Bible Prophecy Web Searches Explode

January 4

Falling Birds Likely Died from Massive Trauma, 500 Dead Blackbirds, Starlings Found in Louisiana Parrish

Kentucky Woman Reports Dozens of Dead Birds in Her Yard

Bees in Freefall as Study Shows Sharp US Decline

January 3

Massive Fish Kill Blankets Arkansas River

January 2

Arkansas' Dead Bird Mystery

Thousands of Birds Fall from the Sky in Beebe

Report: Foot-and-Mouth Spreading in S. Korea

The prophet Hosea connected physical happenings to the moral life of the nation.

"*Listen to the word of the Lord, O sons of Israel, for the Lord has a case against the inhabitants of the land, because there is no faithfulness or kindness or knowledge of God in the land. There is swearing, deception, murder, stealing and adultery. They employ violence, so that bloodshed follows bloodshed. Therefore the land mourns, and everyone who lives in it languishes.* ***Along with the beasts of the field and the birds of the sky, and also the fish of the sea disappear"*** (Hos. 4:1-3).

Jesus reminded satan that man's basic need, more than food, is God's saving word.

"*But He answered and said, It is written, 'man shall not live on bread alone, but on every word that proceeds out of the mouth of God'"* (Matt. 4:4).

There was calamity of fish, birds and animals almost each day of the month. These are symbols of our food. Add to this all the other natural calamities around the world, which have damaged food crops. As I was praying into these incidents of mass extinctions, I was led back to Pharaoh's dreams.

Dream 1:
"*Now it happened at the end of two full years that Pharaoh had a dream, and behold, he was standing by the Nile. And lo, from the Nile there came up seven cows, sleek and fat; and they grazed in the marsh grass. Then behold, seven other cows came up after them from the Nile, ugly and gaunt, and they stood by the other cows on the bank of the Nile. The ugly and gaunt cows ate up the seven sleek and fat cows. Then Pharaoh awoke"* (Gen. 41:1-4).

Dream 2
"*He fell asleep and dreamed a second time; and behold, seven ears of grain came up on a single stalk, plump and good. Then behold,*

seven ears, thin and scorched by the east wind, sprouted up after them. The thin ears swallowed up the seven plump and full ears. Then Pharaoh awoke, and behold, it was a dream" (Gen. 41:5-7).

I pondered on these scriptures and wondered: if the world is now at the end of the years of abundance, then there will be a severe depletion of the food reserves in the world's storehouses.

Unfortunately, the mass extinctions of fish and birds are not dreams, but warning and signs to modern day governments. Will man respond or tragically, continue to ignore the warnings? We have seen in Algeria, Tunisia, Bangladesh, and Egypt, that because the price of food was sky rocketing, rioting occurred. Are the four horsemen fulfilling their last days' assignment, by harming the world food supplies? We will need to wait and watch.[36]

Will You Fill My House Voluntarily?

After the troubling dream about the town crier and the food shortages, I received a second prophetic word from the Lord, which He asked me to deliver. It was a hard prophecy, and I found it particularly hard to deliver: *"If My people will not come to My house to praise, worship, honor, thank Me, and live their lives in accordance with My word, then I will cause them to fill My house for memorial services."* A week later, we had a major accident at a coalmine in New Zealand. Twenty-nine miners died in the mine. The churches around New Zealand began to fill up with memorial services. I cried out in repentance for this nation.

Sometimes one wonders when man will realize the pride in his heart, which prevents him from coming to his maker in repentance and reconciliation. God is patient, but there is an appointed day and time for this generation to face the final judgment of God over the nations. My prayer is that many will get into the Father's house while the gates are still open. For outside those gates is a nightmare, that no man would wish to have.

---

[36] By the 3rd week of February, 2011, Egypt, Algeria, Tunisia, Libya, Bahrain, and Yemen are experiencing serious political riots, demanding old dictatorial governments to resign.

"The Son of Man will send forth His angels, and they will gather out of His kingdom all stumbling blocks, and those who commit lawlessness, and will throw them into the furnace of fire; in that place there will be weeping and gnashing of teeth. Then the righteous will shine forth as the sun in the kingdom of their Father. He who has ears, let him hear" (Matt. 13:41-43).

Even now, dear reader, if you have not yet made your peace with God, you can do it right where you are, at this moment. You can say the "sinner's prayer" below. I pray this often for myself. You can read this prayer by inserting your own name, as you do this by faith in the only Son of God, our Lord Jesus Christ.

"Father God, I (insert your name), come to you in repentance for my sins of rejection of the free gift of salvation in the only name given to men on the earth, the name above all names, the Lord Jesus (Yeshua, is Jesus' Hebrew name, which means "deliverance from sin"). Please forgive me my sins and come Lord Jesus (you can call Him by His Hebrew name, Yeshua) into my heart. I receive you by faith as my Lord and my Savior. Baptize me with your Holy Spirit and make me a child of the Living God, the God of Abraham, Isaac and Jacob our forefathers. Holy Spirit give me a divine thirst for God's word in the holy bible, and teach me to grow as a child of God. In the name of the Lord Jesus (Yeshua), I pray. Amen"

The Church must Preach the Whole Gospel

"For I did not shrink from declaring to you ***the whole purpose of God****"* (Acts 20:27).

Many churches are at a cross roads today about the "whole purpose of God." Paul understood it and was able to teach this with great clarity. Paul was martyred teaching the "whole purpose of God" during his ministry. The early Christians were very clear about living the gospel message, which lead to a large number of Christians accepting martyrdom for what they believed. Today, we find in some parts of the church a theology of compromise. Church leadership are trying to find a gospel that will condone with sin, than face the consequences of holding fast to the hard truth about Lord Jesus' high standards for His people, in purity, holiness and radical pursuit of His Person. In many Christian churches, there is either a tendency to preach less than the whole purposes of God or, in the worst case, to try and improve upon God's purposes to suit the needs of modern man. For example, in Malaysia, a particular church changed the name of God from the Lord Jesus, in their church services, to Allah. Fortunately, the Muslims came against this, as Allah is the name of their god, and they did not want the Christians to take that name and cause confusion. They took the church to court, who ruled in favor of the Muslims. This church was legally required to keep the name of God, as the bible describes Him.

Confusion in the camp is making Christians find a new religion called "Chrislam" that will be a good compromise between Christianity and Islam. We call this Christian modernity![37] Ordained ministers and bishops, in some cases, are embracing gay or lesbian lifestyles. In all of this we will find the church is not in a position to declare the "whole purposes of God" to a generation where a theology of compromise exists in the church.

Even politicians and media personalities are setting themselves up as spiritual leaders these days. They do not have new revelation from God, but greed for power and their undisclosed agendas are promoting their own goals as a form of religion. This is one of the ways, the great deception that Jesus spoke about in Matthew 24:23 and 24 will make its self known. Some Christians will be influenced by these false teachings by respected public figures. They are promoting an unbiblical religion of peace, based on humanistic or eastern ideas of faith and peace.[38]

Many churches are preaching a "watered down" version of the gospel. There is an unbalanced theology about the "love of God"

---

[37] "Chrislam" the new syncretism, is spreading in some western nations: see http://thelastcrusade.org/2010/11/04/chrislam-spreads-throughout-america/

[38] Read Tony Blair's One World Religion move. http://ivarfjeld.wordpress.com/2011/01/27/tony-blair-drags-youth-into-one-world-religion/. Also an interesting article: http://ivarfjeld.wordpress.com/2011/02/22/catholics-in-india-promote-one-world-religion/

without the righteous judgments of God. According to these "new theologians," the days of preaching "brimstone and fire" are over. [39] This version of God is like a comforting, soft pillow. Obviously, their god is not the God who went through the terrible scourging, and nailed to a cross, or, the God who faced the power of the Roman Empire and said, "Yes I am a king." The Lord Jesus, even in the face of death did not compromise on the truth. This falsely comforting theology will fail to keep their followers, in the face of the coming calamities.

"*For then there will be a great tribulation, such as has not occurred since the beginning of the world until now, nor ever will. Unless those days had been cut short, no life would have been saved; but for the sake of the elect those days will be cut short. Then if anyone says to you, 'Behold, here is the Christ,' or 'There He is,' do not believe him. For false Christs and false prophets will arise and will show great signs and wonders, so as to mislead, if possible, even the elect. Behold, I have told you in advance*" (Matt. 24:21-25).

It is true that on the cross we see Christ crucified; this is the love of God, still calling us to salvation. The four horsemen continue their gallop across the nations. God warns through "limited" judgments. Time is indeed running out. The scroll has more seals that will be broken and more severe judgments will follow.

---

[39] The "brimstone and fire" preachers would focus on the terrible consequences of sin. It preached the fear of God into people.

CHAPTER 7

THE FINAL JUDGMENTS

The Fifth Seal is Opened in Heaven by the Lamb

"*When the Lamb broke the fifth seal, I saw underneath the altar the souls of those who had been slain because of the word of God, and because of the testimony which they had maintained; and they cried out with a loud voice, saying, "How long, O Lord, holy and true, will You refrain from judging and avenging our blood on those who dwell on the earth?" And there was given to each of them a white robe; and they were told that they should rest for a little while longer, until the number of their fellow servants and their brethren who were to be killed even as they had been, would be completed also"* (Rev. 6:9-11).

In the above passage, when the sixth seal is broken, John sees the souls under the altar in heaven. These souls are the prophets, slain because they spoke God's word, and the souls of the martyrs, who gave up their lives because they testified to their faith in the Lord Jesus Christ.

These souls are petitioning the Lord to judge and to avenge their martyrdom. However, the Lord tells them that their request must abide a little while longer.

To understand why the Lord does not immediately grant their petition, we must consider what the Lord said to the leadership of the city of Jerusalem in the passage from Matthew quoted below:

"Therefore, behold, I am sending you prophets and wise men and scribes; some of them you will kill and crucify, and some of them you will scourge in your synagogues, and persecute from city to city, so that upon you may fall the guilt of all the righteous bloodshed on earth, from the blood of righteous Abel to the blood of Zechariah, the son of Berechiah, whom you murdered between the temple and the altar. Truly I say to you, all these things will come upon this generation. Jerusalem, Jerusalem, who kills the prophets and stones those who are sent to her! How often I wanted to gather your children together, the way a hen gathers her chicks under her wings, and you were unwilling. "Behold, your house is being left to you desolate! "For I say to you, from now on you will not see Me until you say, 'blessed is He who comes in the name of the lord!'" (Matt. 23:34-36).

In this passage from Matthew, the Lord Jesus speaks about the same thing: the blood of prophets and martyrs, crying out to God in heaven, which John also sees in the Revelation, after the fifth seal is broken. **The Lord Jesus connects this cry or petition of the prophets and martyrs to the fate of the city of Jerusalem.**

To get to a deeper understanding, since it relates to prophetic intercession, we will need to see both of these passages taken together, as well as to understand the significance of the altars in the temple.

The Two Altars and the Two Temples

The Old Testament temple had two altars. The outer court had the bronze altar, where the burnt offering was sacrificed. Inside of the temple was the altar of incense. This incense altar was in the Holy Place, and its fragrant cloud of smoke was used by the priest to permeate into the Holy of Holies and to cover the Mercy Seat. The priest could not see the Sh'chinah (Shekinah) Glory of God residing over the Mercy Seat. If the cloud of incense did not hide God's Face from the priest, he would die, while performing his priestly duties. This incense represents the prayers and intercessions of God's people, which are like a shield to us and to those for whom we stand in the gap. Just as the priest played an intercessory role for Israel, so we, who are the priests of the New Covenant, stand in the place of intercession, between a Holy God and a sinful people.

The altar of incense in the temple of heaven: John sees the souls of the martyrs, who are crying out for God to avenge their blood. These are intercessory prayers for justice, which the incense symbolizes. They gave their lives on earth, and their blood cries from the ground, while their souls cry out in heaven. They must wait until the full number of martyrs is complete, before God will execute justice and avenge their blood.

The altar of sacrifice in the temple on the earth: The bronze altar of sacrifice symbolizes the blood of the martyrs offered on earth. The Lord Jesus summarized all the martyrs of history, by citing the blood of righteous Abel to the blood of Zechariah, son of Berechiah, and this extends to all those who were killed because of their faith in the one true God.

Jesus and John hear the cry of the blood of the martyrs in heaven's intercession, and they know that the cry is for the punishment of the cities and nations on earth, which had shed innocent blood. If the cities and nations of the earth do not repent of their deeds and rejection of the Messiah, they will be judged.

However, the Lord in His great mercy, asks the martyrs to wait a little longer, saying that the full number of martyrs must be complete. In saying this, He is giving the nations and cities of the earth more time to repent. He is delaying the judgment, at the expense of the future generations of martyrs, whose blood will fill up the cup of His wrath.

When will we cry out for our cities? If the church does not, their fate is sealed as was Jerusalem's fate. The blood of the martyrs of your city calls for God's justice. Is there no one to cry out to the Father for His mercy on your cities? The greatest failure of the church would be to hear the Lord Jesus tell her, "*I looked for a man and found none.*"

Understanding the Doctrine of National Repentance

When saints are martyred on earth, their shed blood demands God's justice. Those who shed their blood will need to pay with their own blood, unless their wicked deeds are atoned for by the blood of Yeshua. This must be combined with repentance and confession of their sin. **However, someone must stand in the gap and intercede for the guilty, or the Holy Spirit will not intervene and convict them, and they will be destroyed by God's wrath**. The souls in heaven are crying out for justice. God's justice can satisfied in one of two ways.

- The Lord can destroy those who destroyed them; or
- The Lord can pay for this crime with the blood of His Son, and with the repentance of the murderers.

In the same way, when a nation or a city is wicked, there are only two ways for God's justice to be satisfied:

- The Lord can destroy the city with fire or warfare; or
- The city can repent, and appropriate the punishment that fell upon Yeshua to their own guilt, and God's wrath will be satisfied at the expense of His Son, without having to destroy the city.

"And the word of the Lord came to me, saying, 'son of man, speak to the sons of your people and say to them, If I bring a sword upon a land, and the people of the land take one man from among them and make him their watchman, and he sees the sword coming upon the land and blows on the trumpet and warns the people then he who hears the sound of the trumpet and does not take warning, and a sword comes and takes him away, his blood will be on his own head. He heard the sound of the trumpet but did not take warning; his blood will be on himself. But had he taken warning, he would

have delivered his life. But if the watchman sees the sword coming and does not blow the trumpet and the people are not warned, and a sword comes and takes a person from them, he is taken away in his iniquity; but his blood I will require from the watchman's hand'" (Ezek.33:1-6)

In this chilling passage from Ezekiel, the church, the prophetic voice in the city, must call the city to repentance. To fail to do so will mean that on Judgment Day, the church who did not warn the wicked will be held accountable for their blood.

The word "avenge" in the Greek indicates a legal action (see Luke18:3,5). The breaking of the fifth seal, what the martyred saints are praying for, is not "an eye for an eye or a tooth for a tooth," but that the Lord Jesus would initiate the process to legally take back the earth.

In John's Revelation, as mentioned earlier, the breaking of each seal of the scroll in heaven initiates the end-time protocol. John is awakening the church to fervent prayer for the cities of the world.

Since the beginning of the 20th century, over 45 million Christians were martyred for their faith in the Lord Jesus Christ (see David Barrett, *World Christian Encyclopedia*, 2001[40]). There is not enough prayer going up for our cities, nor for the persecuted church in most church services. It is not my intention to criticize the church, but there is an unholy "preoccupation" with ourselves and with "our church," that excludes and makes no room for God's agenda for the return of His Son. There are more sermons and more books available on how to grow financially through tithing, than on calling the church to pray for its persecuted brothers and sisters in different parts of the world.

Many of my intercessor friends who were part of the International Fellowship of Intercessors, and respected authors of several books on prophetic intercession, like Kjell Sjoberg of Sweden and Johannes Facius of Germany, labored in many years of prayer for the persecuted church in communist Europe. Kjell and Johannes made numerous "prayer missions" into nations which were, at that time,

---

[40] The data is available online: http://www.bible.ca/global-religion-statistics-world-christian-encyclopedia.htm

"behind the iron curtain," even at the risk of their lives. Johannes had a physical battle with satan in the U.S.S.R., which caused a partial loss of his memory. The Lord did restore it fully, and he now continues to serves the Lord through an international prophetic bible teaching ministry. These humble and faithful servants of the Lord (Kjell has since gone to be with the Lord), have seen the fruits of their intercession and spiritual warfare. Many nations behind the "Iron Curtain," including Russia, were liberated from communism. Glory be to God, who gives us the victory through Christ (1 Cor. 15:57).

Johannes Facius, in his tehth newsletter, "Voice of a Watchman" has this to say:

"Is it any wonder that Christianity in our part of the world is on the retreat everywhere? We have become like a sick and dying patient. More and more church buildings here in Europe are now for sale. In my birth town of Copenhagen there are at present ten Lutheran churches for sale. Many others have already been sold and are now taken over by fitness centers. In one of them, it is written over the altar: "God does not live here anymore." Islam is on the move and gaining ground, as are other religions like Buddhism. Here in Germany, the Dalai Lama has recently assembled a crowd of 10,000 people every night for a week. No Christian leader or evangelist is able to do that anymore.

"West European Christianity has become a weak bowl of humanistic sweet soup and is by-and- large in danger of becoming the church of Laodicea, of which the Lord said: '*I could wish you were cold or hot. So then, because you are lukewarm, and neither cold nor hot, I will vomit you out of my mouth*' (Rev. 3:15-16).

"Thank God for those who stand up and speak up. Thank God for men like the Editor of Charisma Magazine, Brother J. Lee Grady. I was also encouraged by some statements made by Rev. Thomas E.Trask, the outgoing superintendent of the Assemblies of God in the USA. This is what he recently said in his resignation letter: 'I have taken time to seek the Lord as to His direction.' During his farewell speech, Trask encouraged the largest Pentecostal denomination in the world to uphold righteousness and reject all forms of sin. 'What many churches are offering America is a new religion that guarantees no hell and requires no holiness,' Trask said. 'It is a limp, spineless Christianity that does not confront sin for fear of being judgmental. It is an impotent gospel that tells people everything is OK.'

"Paul, the Apostle, said '*...For the time will come when men will not put up with sound doctrine. Instead, to suit their own desires, they will gather around them a great number of teachers to say what their itching ears want to hear.*' (2 Tim. 4:3 NIV)

"Well, the time is here and now. We are right in the midst of it. May the Lord help and protect His people." [41]

Apart from praying, we need to direct our support for these persecuted churches. There are numerous ways to show our love and concern. Monetary support is one way, as well as sending clothes, building schools, improving sanitation needs, providing clean drinking water, teaching them trade skills, and even visiting them to encourage them. These are some ways to show our love and support for these great saints. Praying for their needs is a great way to support our persecuted brethren. Visiting the persecuted church in some nations may be difficult and dangerous. However, if God calls you to visit them, that would be a wonderful blessing to them.

Recently I was ministering in a church in the southern part of the North Island of New Zealand. As I prayed over a couple, I had a strange vision. I saw huge ship come out from the sea and on to the beach. It finally came to a halt between the couple. I had no word of knowledge as to what this vision meant. I shared this strange vision with the couple. It struck no immediate bells. In fact, the good gentleman joked that he had no plans to take his wife for a cruise. We laughed and left. At the end of the ministry service, as I left the church building, the couple was waiting for me outside the church

[41] News letter no. 10 "Voice of a Watchman" by Johannes Facius. In 1985 Johannes became the international director of IFI and led this global prayer movement until the end of 1999. I have had the privilege of being a member of the IFI from 1989 to 2001, and served under Johannes' able leadership. Johannes is the author of several books on prophetic intercession. Credit: Johannes Facius http://www.ifi-d.de/index.php?id=41&L=1 (no copyright restrictions).

building, in the foyer. The lady said she wanted to tell me something. The vision did speak to them.

Here is their story: The gentleman (her husband), was an architect by profession, now retired. He had worked for "Habitat," a company that did low cost housing from local materials in poor nations. Their daughter was a missionary in Vietnam. She often wrote to her father, asking him to come to Vietnam, where she worked among the poor. She wanted him to teach the local people to build low cost housing. The good man used to read his daughters letters, but could not make up his mind to go to Vietnam, a nation where the persecuted church exits. Whenever his wife would ask him when he would make his decision to help his daughter, his reply was "We will wait till the ship comes home." The vision helped him to make up his mind about visiting his daughter and helping the persecuted church in Vietnam with his skills as an architect. I was glad to be of some help here through the vision that the Lord gave me for them.

Evaluate where you are giving your tithes and offerings, and ask the Lord to help you to come into alignment with His end time plans. The persecuted church needs much support to care for the harvest in some of the world's worst anti-Christian environments. Some of the pastors and members who serve in the ministry of the National Days of Prayer in New Zealand, take part time employment, in order to financially support the persecuted churches from their wages. My friend and fellow laborer, Pastor Warren Lyons, the founder president of The National Days of Prayer, supports the pastors in Myanmar, Indonesia, Vietnam and Cambodia, toward their pastoral service needs. Often under the threat of his own life, he travels, reaching out and meeting their needs for bibles and Christian teaching. If God is calling you to financially support this work, please contact me by e-mail, full details can be found at the end of the book.

The Sixth Seal is Opened in Heaven by the Lamb

"I looked when He broke the sixth seal, and there was a great earthquake; and the sun became black as sackcloth made of hair, and the whole moon became like blood; and the stars of the sky fell

to the earth, as a fig tree casts its unripe figs when shaken by a great wind. The sky was split apart like a scroll when it is rolled up, and every mountain and island were moved out of their places. Then the kings of the earth and the great men and the commanders and the rich and the strong and every slave and free man hid themselves in the caves and among the rocks of the mountains; and they said to the mountains and to the rocks, 'Fall on us and hide us from the presence of Him who sits on the throne, and from the wrath of the Lamb; for the great day of their wrath has come, and who is able to stand?'" (Rev. 6:12-17).

Chapter 6 ends with the opening of the sixth seal. Heaven has moved away from its "partial judgments," and the nations now stand before the Lord to face His righteous judgment. The sixth seal marks an awesome dramatic time in human history.

The Lord Jesus walked the earth, and He taught about the kingdom of God as both something that is within us, in the here and now and as a future event that was still to come. In His first coming, He established His kingdom rule in the hearts of His followers, His church. This is a spiritual concept. One can look at the church as a territory within the enemy's camp, which belongs to the kingdom of God. While His disciples were in a geographical kingdom of darkness, their allegiance was not to the rulers of darkness, but to King Jesus. This is the "kingdom of God within" concept.

In His second coming, the coming of the kingdom of heaven on the earth is a geographical concept. John sees the future event of the kingdom of God, when he says that the kingdoms of the earth have become the kingdoms of our Lord and of His Christ (see Rev.11:15). In the Greek, the word for kingdom is "Basileia," denoting sovereignty, royal power, and dominion. It further denotes the territory or people over whom a king rules (see Vine's Expository Dictionary). Thus, the word "kingdom" indicates both power and government. It also signifies the extent of both the territory and ruling authority of the kingdom. Hence, the basic meaning of kingdom involves three things: a ruler, a people who are ruled, and a territory over which they are ruled. The kingdom of heaven is not just a theological concept limited to a realm called heaven. It is the government of God, which takes over the earth and establishes the rule of God on the earth. It will involve territory (all the earth), people and a ruler, who is the Lord Jesus.

Some Christian prophets of today have actually experienced heaven as a physical place. There is not enough science to prove this, but our science has not advanced enough to know everything in God's awesome universe. The great scientist Einstein died of frustration, trying to prove the "the theory of everything," in relation to how the entire universe is all linked together.

As the heavenly kingdom of God moves towards the earth, surely, we will experience its effects, magnetically, gravitationally, and in other physical phenomena. This will cause huge upheavals on the earth, causing earthquakes, tsunamis, the very shaking of the earth.

"The earth reels to and fro like a drunkard and it totters like a shack, for its transgression is heavy upon it, and it will fall, never to rise again" (Isa.24:20).

To some degree, there has been a rehearsal in smaller measures of some of these events, as though it is warning us of that which is to come.

- **Earthquakes that move nations:** Consider this bit of news: Wednesday, July 22, 2009: A massive earthquake last week has brought New Zealand closer to Australia, scientists say. The 7.8 magnitude quake in the Tasman Sea has *expanded* New Zealand's South Island westwards by about 30cm (12in). The earthquake of Revelation 6:12 will be far mightier than this one reported. The Greek work "seismos" is translated as earthquake, but literally means, "shaking," and can be applied to land, sea or air. People on the earth will experience it as both the land and sky shaking. Haggai 2:6 and 7 declares that the Lord will shake the earth, the sea and the dry land, and He will shake the nations. God sends precursors to warn us.
- **The Sun will be darkened:** Timothy Dwight recorded in his journal, "The 19th of May, 1780, was a remarkable dark day.

The morning was clear and pleasant, but about eight o'clock there was observed an uncommon appearance in the sun. There were no clouds, but the air was thick, having a smoky appearance, and the sun shone with a pale, yellowish hue, but kept growing darker and darker, until it was hid from sight. There was midnight darkness at noonday. Candles were lighted in many houses; the birds were silent and disappeared, and the fowls retired to roost. ... A very general opinion prevailed, that the Day of Judgment was at hand." This was quoted in Connecticut Historical Collections, compiled by John Warner Barber, 2nd edition, New Haven: Durrie & Peck and J.W. Barber, 1836, p. 403.

- **Stars falling out of heaven and on to the earth:** Here's another bit of news that may look like stars falling out of the sky and on to the earth. **EMP Attacks**—EMP stands for electromagnetic pulse, and according to a U.S. Threat Assessment Commission, it presents that rarest of all security risks: an attack that can deliver a devastating blow to the U.S. military, and kill tens of millions of U.S. civilians in the bargain. The idea is as simple as it is horrific. If a nuclear device is detonated in the air, the explosion sends out an EMP, which shuts down all electrical systems in the broad, multistate region below the explosion. In the modern United States, hardly anything works when the electronics go down. No transportation, no heat, no lighting, no communications, nothing that modern man takes for granted in his daily walk of life, would exist. Hospitals cannot function. Food, medical and other supplies are immobilized. In a matter of days, masses of people start dying. The EMP Commission ominously warned that China and Russia have considered limited nuclear attack options that, unlike their Cold War plans, employ EMP as the primary or sole means of attack. Indeed, Russian, Chinese, and Iranian military writings abound with references to EMP strikes against the United States.

James Jay Carafano, Ph.D., and Richard Weitz, Ph.D., two of the country's preeminent national security experts explain how to prevent the worst. Their report says: Most Americans, whether members of the public or politicians in Congress, ignore or are unaware of the very real threat of an electromagnetic pulse (EMP) attack. A nuclear device detonated high in the atmosphere above the American mainland can easily disable the country's electrical grid, shutting down nearly all communications, transportation, and service systems. Overnight, daily life as Americans know it, will be a thing of the past. There are ways to prevent devastation from an EMP, and the U.S. must invest in them now before it is too late. Not even a global humanitarian effort would be enough to keep hundreds of millions of Americans from death by starvation, exposure, or lack of medicine. Nor would the catastrophe stop at U.S. borders. Most of Canada would be devastated, too, as its infrastructure is integrated with the U.S. power grid. Without the American economic engine, the world economy would quickly collapse. Much of the world's intellectual brain power (half of it is in the United States) would be lost as well. Earth would most likely recede into the "new" Dark Ages. [42]

- **The world is on the brink of a nuclear war**

  A news report, December 17, 2010. North Korea warned that another war with South Korea would involve nuclear weapons. If war breaks out, it will lead to nuclear warfare and *not be limited to the Korean peninsula.* We live daily on the thin ice of a sudden break out of a nuclear war. When this happens, EMP will look like stars falling out from heaven on earth. Current tension in Middle East and in Asia has both the direct and indirect potential to lead to nuclear war. More nations have nuclear weapons today than ten years ago. Apart from Advanced Western Nations, now North Korea, Syria and Iran have developed long range ICBM capable of carrying nuclear warheads. A classified Pentagon study in 2002, estimated that the deaths could be in the order of 12 million people, should a nuclear war break out between two nations. In another news report, Al-Qaida is on the verge of producing "radioactive" weapons. Today, the raw materials for making

[42] To read the entire article on the web: http://www.heritage.org/research/reports/2010/11/emp-attacks-what-the-us-must-do-now

> these weapons are available in the weapons market. Nuclear terrorism, in the words of the President of the U.S.A., Barack Obama, is the "single biggest threat" to international security, with the potential to effect "extraordinary loss of life."[43]

This is obviously going to cause global panic the proportions of which are unprecedented. Governments will collapse world-wide throwing nations into chaos. With the foreknowledge of these coming events, the church needs to prepare for a time such as this.

The church needs to prepare itself for the greatest harvest of all times. Church disciple-making programs should prepare its congregations to understand what is to come, and how to respond with faith and bring in the harvest. We will have to give account to the Lord for the stewardship of the people who will come knocking on the doors of our churches and homes, in the days that will befall the earth. Can we say that our churches are preparing to handle the coming harvest? The Lord Jesus gave this warning following His discourse on the end-times, in Matthew 24.

"For I was hungry, and you gave Me something to eat; I was thirsty, and you gave Me something to drink; I was a stranger, and you invited Me in; naked, and you clothed Me; I was sick, and you visited Me; I was in prison, and you came to Me.' Then the righteous will answer Him, 'Lord, when did we see You hungry, and feed You, or thirsty, and give You something to drink? 'And when did we see You a stranger, and invite You in, or naked, and clothe You? 'When did we see You sick, or in prison, and come to You?' The King will answer and say to them, 'Truly I say to you, to the extent that you did it to one of these brothers of Mine, even the least of them, you did it to Me.' Then He will also say to those on His left, 'Depart from Me, accursed ones, into the eternal fire which has been prepared for the devil and his angels; for I was hungry, and you gave Me nothing to eat; I was thirsty, and you gave Me nothing to drink; I was a stranger, and you did not invite Me in; naked, and you did not

[43] Article was on "Al-Qaida on brink of nuclear bomb" http://www.vancouversun.com/news/Qai...#ixzz1ClQs3zvR

clothe Me; sick, and in prison, and you did not visit Me.' Then they themselves also will answer, 'Lord, when did we see You hungry, or thirsty, or a stranger, or naked, or sick, or in prison, and did not take care of You?' Then He will answer them, 'Truly I say to you, to the extent that you did not do it to one of the least of these, you did not do it to Me.' (Matt. 25:35-45)."

These scriptures indicate the situation of people in the world who cannot run to any place to save themselves (see Rev. 6:6). They will come knocking on our doors for help in their bewilderment. The faith of most Christians will be tested. The testing will be trusting God for His provisions, while helping those in need, drawing them to salvation, and making them disciples of the Lord. There is a severe warning not to neglect Christian charity and hospitality in this hour, and this time of world chaos.

It is not just enough to gather in the harvest, but also to prepare them for the marriage feast of the Lamb. This is the church's work, to make *disciples* of all nations. The church needs to consider many ways to handle the future coming events. The church of the millennium is the church that reigns and rules with Christ. Hence, the church needs to plan for the futuristic church of the Christ millennium.

Several areas come to mind regarding discipleship. Christians who work in various societal infrastructures like education, business, government, trade and commerce, media and entertainment, the arts, and everything that touches our day-to-day living, need to be trained in the Kingdom of God principles. For example, how does a Christian business man put into effect in his personal life, his family life and his business life, the principles of the kingdom of God? In the Millennium, to reign and rule with Christ will mean putting on the mind and heart of Christ, and the principles of His Kingdom, into every sphere of his life. If He can trust you now to live by Kingdom principles in your earthly affairs, He will find you trustworthy to rule and reign with Him when His Kingdom comes to earth.

Church leadership will therefore need to train pastors, house-groups, church-groups, market-Christians, teachers or professors, and just about every group, how to apply Kingdom principles in these diverse fields. It is not going to be good enough just to have bible studies or evangelistic studies and stop there. It takes more than just biblical knowledge to run an entire nation. As Jesus said, the kingdom of God is at hand, which means that Christians must learn how to govern, and exercise authority under Messiah's Lordship now. Even if you are not in an earthly position of authority, you are nevertheless being trained to rule and reign with the Lord in every sphere of influence in your life.

In this way, the Spirit of God is preparing the bride to get ready for the return of the Lord Jesus to earth. The oil in the lamp of the ten virgins is not just the oil that burns the sanctuary lamps of the church. The oil is the anointing, which can only be obtained through intimate communion with the Holy Spirit. Individuals who have cultivated transparency and intimacy in the Secret Place of the Most High, will qualify to receive authority in the coming Millennium reign of the kingdom of God on earth. The anointing in this corporate body of wise servants of the Lord, to them will be granted to govern in world economics, education, politics, and all of the Millennial society's infrastructures. These ones are like the five wise virgins in the Lord Jesus' parable of the ten virgins. The oil of the anointing stored up by intimacy with the Lord, will become the oil of anointing to rule and reign with Him in His mellenium reign on the earth.

Leaders who procrastinate in preparing their flock for the Millennium, are like the five foolish virgins, who did not take the time or trouble to obtain oil in their lamps. They missed the Wedding Supper of the Lamb. Five out of ten were not ready for His coming. Is the Lord warning us that fifty percent of the church is not ready? The Lord is moving among His lamp stands, to see if they are ready. There is coming an irrevocable moment when He will check for light (understanding) and oil (anointing) for the last time. The Lord's warnings to the seven churches in the book of Revelation are a warning to the church even today, to get ready for His return.

Intercession

God has begun His judgments on the nations. We worship our God with the great Hallelujah, in heaven, for the kingdoms of the earth have become the kingdom of our Lord and of His Christ. The church needs to seek the Lord's guidance, with intense prayer, as to how to prepare for His coming. Intercession should precede planning. "God, teach us how to plan for the coming harvest and how to reign and rule with You in the age to come," should be the cry of our hearts today.

Property values in most nations have fallen. Churches need to use this opportunity to invest in property, for the sake of the harvest. This is one example of planning for the future. Christian businessmen need to invest in the coming harvest; not only looking to what the ROI (return on investment) on your money would get you in a secular world, which will fall anyway, but look for the ROI in the kingdom of heaven on earth. We must pray that the church is blessed with people like Daniel and Joseph, who have wisdom to build a nation on the God's kingdom principles. While deep darkness covers the earth, there must be the sound of joy wafting from the house of the Lord.

"*Arise, shine; for your light has come, and the glory of the Lord has risen upon you. For behold, darkness will cover the earth and deep darkness the peoples; but the Lord will rise upon you and His glory will appear upon you. Nations will come to your light, and kings to the brightness of your rising.*" (Isa. 60:1-3).

In the context of all that has been said here about the opening of the sixth seal, let us deeply consider the words of Isaiah: The Lord promises that even to the end of man's days, HIS mercy never ends, even at the eleventh hour.

"*The afflicted and needy are seeking water, but there is none, and their tongue is parched with thirst; I, the Lord, will answer them Myself, as the God of Israel I will not forsake them.*" (Isa. 41:17).

Intercession for the Remnant of the Tribes of Israel

"*And I heard the number of those who were sealed, one hundred and forty-four thousand sealed from every tribe of the sons of Israel.*" (Rev. 7:4).

Revelation 7 describes a special category of people, those to whom the Spirit of God turns our attention. These are the Jewish people from the tribes of Israel, who will be converted to the Messiah, Lord Yeshua, during the time of "partial judgments" of chapter six. This category of people the remnant of Israel the restored ones, as promised by God's covenant promise to Israel.

Paul testifies to this covenant promise when he stands trial before King Agrippa:

The promise to which our twelve tribes hope to attain, as they earnestly serve God night and day. And for this hope, O King, I am being accused by Jews." (Acts 26:7).

Paul's letter to the church in Rome again speaks of "the remnant" of Israel that will accept Messiah Yeshua as the Lord and Savior.

"*I say then, **God has not rejected His people**, has He? May it never be! For I too am an Israelite, a descendant of Abraham, of the tribe of Benjamin. **God has not rejected His people whom He foreknew**. Or, do you not know what the Scripture says in the passage about Elijah, how he pleads with God against Israel? Lord, they have killed your prophets, they have torn down your altars, and I alone am left, and they are seeking my life. But what is the divine response to Him? I have kept for Myself seven thousand men who have not bowed the knee to Baal. In the same way then, there has also come to be **at the present time a remnant according to God's gracious choice.**"* (Rom. 11:1-5).

In a time of chaos, confusion, and increased lawlessness, Israel will be under much stress from her enemies around her, trying to persecute her. The prophet Elijah, in order to save his life from the wicked Jezebel, hid himself from the fear of death. This remnant, from every tribe of Israel has a purpose, which we see in Revelation 7:9-12. They are the ones that will lead the multitude from every tribe, tongue and the nations of the world, to God. This remnant of Messianic Jewish leaders who are sealed by God for a very specific role in the end times, just as the prophets that were hidden during the Jezebel rage that made Elijah flee for his own life. **This is the "new and final Exodus." Just as Moses led the children of God out of Egypt, so this "Moses Company" (the one-hundred and forty-four thousand Jews) will lead His captive Bride out of slavery, out of the grip of the antichrist, and into the Kingdom Age, her Promised Land.** The exile of Israel becomes the hope for the nations. Heavens gates close after this final call, heralding by the trumpet blast, the last and final seal is ready to be broken in heaven.

The church must pray for this remnant of Israel. We must pray for this special group of people, who will respond to the Spirit of God, and will carry out her God-given role of leading the people to God. We must pray for them to be courageous in the face of persecution, and to lead this remnant from the nations to God. This is their end-time assignment, in the face of tremendous persecution and hardship, and they will do it. **The church needs to obey the word of God here and purge itself of any anti-Semitic attitude. These are precious people of God carrying out one of the most insidious tasks in the end times**. The church must hold them dearly in their hearts with much love and supplication for them. I do believe their task of collecting the final harvest from the earth, before the seventh seal is broken may be just after the rapture of the church, but before the Lord returns to the earth to set up His kingdom rule in the millennium. If we understand this situation, our hearts will go out to these awesome obedient servants of God who loved not their lives for the sake of the love of the Messiah, and to do His bidding.

Silence in Heaven before the Seventh Seal is Opened.

The opening of the first six seals in heaven brings about much activity on earth. Earthquakes, famines, floods, crop damage, wars, and much loss of life are some of the devastating consequences.

These are just partial judgments of God prior to the opening of the seventh seal.

There is a silence in heaven for about half an hour before the opening of the seventh seal.

"*When the Lamb broke the seventh seal, there was silence in heaven for about half an hour.*" (Rev. 8:1).

Jewish tradition speaks about angelic silence in heaven, in ***Jewish liturgy.*** This tradition says that when Israel prays, the angels go silent, and God's judgments are about to fall.

In Revelation 4, all of creation in heaven praises God day and night. However, there is an interruption in the "heavenly liturgy," to allow the intercessory prayers of God's people on earth to be heard. It is as if heaven waits in silence to hear the voice of the bride of the Lamb of God. When all the angels and saints in heaven, and the Lamb, hear the voice of the bride, all action for final judgment begins.

Liturgy has a place in heaven's worship, warfare and intercession

In the Temple period, many special readings and liturgical meditations developed as an expression of Jewish worship. The Psalms, the longest book of the Bible, in essence is a book of liturgy. Hundreds of years before the coming of the Lord Yeshua, certain prayers were common. For example, the *"Shema"* (Deut. 6:4-9), the *"Amidah"* (18 benedictions, later changed to 19 because of the early Messianic Jews) and some scholars even believe such prayers as the *"Kaddish"* and the *"Aleinu"* were in fact used before the first century as well.

These prayers were all part of Jewish liturgy, which the Lord Jesus was very familiar with, as part of His Jewish tradition. We are aware of His teaching ministry, but can you imagine Him being called upon as a respected rabbi (or good Teacher – Luke18:18) if He did not enter into the liturgical worship of His day? For Him, it must have been a beautiful expression of praise to the Father. When the Lord Yeshua was asked one day, what was the greatest of all the commandments of Torah, He quoted the *Shema.* This prayer is the heart of the liturgical worship for the Jew (Mark 12:28-34). Then again, the "Lord's Prayer" (Matthew 6:9-13), is also a summary of many Jewish prayers.

- "Our Father" – 5th Benediction of the *Amidah*
- "Hallowed Be Thy Name" – the *Kaddish*
- "Thy Kingdom Come" – the *Kaddish*
- "Thy Will Be Done" – the *Kaddish*
- "Give Us Our Daily Bread" – 9th Benediction of the *Amidah*
- "Forgive Us Our Debts" – 6th Benediction of the *Amidah*

It seems clear that the Lord Yeshua not only worshipped through the liturgy but also quoted it in the course of His teaching!

The *Amidah* is the central prayer four Jewish liturgical services:
The *shacharit* (morning prayer),
The *mincha* (afternoon),
The *maariv* (evening), and
The *mussaf* (additional).

The word *Amidah* literally means standing. The prayer is recited while standing. It is also known as *Shemoneh Esrei*, meaning eighteen, because it originally consisted of eighteen blessings, and as *tefilah* (prayer) because it is the most important Jewish prayer.

The obligation to pray three times a day, which was established by Ezra and codified in the Talmud (*Berakhot* 26b), is fulfilled by reciting the *Amidah.* [44]

One can easily see how the traditional Christian churches continued with the practice of Jewish liturgical prayer, which is called the "Divine Office." Liturgical prayer meant for the "hours of the day," morning, afternoon and evening prayers. I enjoyed singing the

[44] Schoenberg, Shira. see Article, "The *Amidah*," in the Jewish Virtual Library, (for an excellent, deeper reading), A division of the American-Israeli Cooperative Enterprise. Copyright American-Israeli Cooperative Enterprise, Reprinted with permission. http://www.jewishvirtuallibrary.org/jsource/Judaism/amidah.html

psalms and praise songs at 3:00 am every morning with the Catholic Franciscan monks at Mount Calvary, while I was in Israel on a prayer mission.

Liturgy, as we experience it in traditional churches, has its origin in Jewish synagogue services. Take for example, the Passover service that observed in each Jewish home. The Lord Jesus, while having the last Passover supper with His disciples, instructed them to repeat the ceremony (liturgy) of breaking bread and sharing wine to remember Him after He was gone. The disciples, attempting to emulate this ceremony, established the liturgy into the New Testament times. The Lord Jesus asked the disciples to continue to remember Him, and to remember His sacrifice for us, by continuing to have the "Lord's Supper." ***It is a remembrance until He returns***. Although the liturgical procedure may differ somewhat from church to church, the significance remains the same: Christ died on the cross for our sins, Christ has risen, and Christ will come again in all of His power and glory as the King of kings, to claim His kingdom and His bride on earth.

In trying to understand how the word liturgy came into the Christian vocabulary, we will need look at the Greek word *"leitourgia,"* which means a public duty, a service to the state undertaken by a citizen. The Greek word is made up of two root woods: *leitos* (from *leos = laos*, people) meaning *public,* and *ergo* meaning, *to do*. Thus, the word *leitourgos* means, "a man who performs a public duty." One would wonder, why then, is the word tied to a "holy service" in the church today? A "public servant" was a Roman *lictor;* and *leitourgeo*, "to do such a duty."

The Septuagint, (abbreviated as the "Seventy" in Latin, "LXX,") is the Greek translation of the Jewish Scriptures. These were the Jewish Scriptures used by the Jews who scattered across the Roman empire. They used it for their "public service" in the temple. This gave the word "liturgy" a place in the service of the priests, a ritual service, and lending meaning to the "liturgical service" in the temple. It has come to mean an established pattern for public worship. In Christian circles, liturgy meant the public official service of the Church, which corresponded to the official service of the Temple in the Old Testament[45]

Liturgical churches are generally referred to as those with an emphasis on a traditional practice, and where the words for worship and service are formally written out. This is in contrast to a less structured service style, as practiced by many of the Catholic, Protestant, and Orthodox Christian churches.

Most of these traditional churches follow a "liturgical service." At the heart of church liturgy is the common public prayer. "Lauds" or Morning Prayer, then there are the "Vespers," or at the end of the day prayer, "Vigils," or prayer at night. Then there are the "Terce", "Sext" and "None" which is prayer at the 3rd hour, 6th hour and 9th hour.[46] However, these are structured prayers, based in the word of God. Some churches have prayers based on tradition that seems to have come down from the early church soon after church began to spread into Europe. These early Christians met regularly to celebrate the Lord's Supper, to learn from the teachings of the apostles called the "Didache" or The Teaching of the Twelve Apostles.

During the first few centuries of Christianity, the early Christians were Jewish converts and some Gentile converts. A great variety of prayer patterns flourished in the Christian Churches throughout the Roman Empire and beyond. Common to all these patterns, though, were the meetings for common prayer in the morning, the evening, and occasionally during the night. The powerful symbols of light and dark, the rising and setting of the sun, came to be an integral part of these prayer services. Morning prayers focused on the Risen Messiah, evening prayers on the continual need for forgiveness and protection from cosmic forces, and night prayers, on the coming of the Messiah at the end of time. This should inspire us to follow the book of Revelation in worship, warfare and intercession in the march towards preparatory prayer for the return of the Lord.

---

[45] For more details see Catholic online http://www.catholic.org/encyclopedia/view.php?id=7159; for a very detailed information on the Septuagint, see the web site "All about God," http://www.septuagint.net/

[46] Wikipedia, the Free online encyclopedia, "Liturgy of the Hours." http://en.wikipedia.org/wiki/Liturgy_of_the_Hours

Liturgy can lose its significance, power and grace if it becomes a ritual. A thin line separates liturgy from ritualism. The danger of repetitive prayer is that it can become a ritual without meaning. When liturgical services focus on the outward performance of the ritual, rather than heartfelt prayer to God our Father, through the Lord Jesus, it runs a danger of falling into the trap that Paul wrote: holding to an outward form of religion, but denying its power (see 2Tim.3:5). I believe that even within the context of a structured liturgical service, there should be opportunities for the people to move in the revelation and the power of the Spirit of God. This brings life, for the Spirit is life, the flesh is of no avail (see John 6:63).

There are some liturgical services based on the book of Revelation,[47] which could draw the church into worship and intercession with heaven, for return of the Lord Jesus. In fact, the church should pause at each movement in the liturgy to allow God's people to enter into the Spirit's revelation to freely worship and intercede. Take for example, the liturgy of repentance. Often the focus is only on personal sinfulness and repentance. There is always a place for growing in personal holiness. However, the prayer begins with "calling to mind our sins." At moments like these, the church should remember the sins of previous generations, of the founding fathers of the city or nation, the sins of the current generation, and the sins of our national culture. We must ask for the blood of the Lord Jesus to cleanse our city, our nation, from all unrighteousness. We can thank God that in His judgments there is a place for mercy, and that He has been so gracious to give us time to repent for the land.

The communion service in the liturgy is a time for the bride, God's people, to passionately express her longing for the return of her Lord. "Come Lord Jesus, we long for Your return, do not delay." **This freedom of expression of love to the Lord Jesus, is what heaven is waiting to hear from the bride on the earth.** Liturgy should assist and even promote the free expression of love for the

---

[47] Hahn Scott, The Lamb's Supper: The Mass as Heaven on Earth, Bantam Dell Pub Group, 1999. Catholic theologian Dr. Scott Hahn thinks that many Catholic worshippers receive communion without ever considering its links to the end of the world, the Apocalypse, the Second Coming of our Lord Jesus Christ to the earth and its implications in our life and ministry to the world.

Lord, not stifle it. This can only happen when each member in the pew and the priest at the altar have hearts throbbing passionately for the Lord's return. This serves as an example of how traditional churches can use the power and grace of liturgical movements in worship and intercession, joining heaven, as it prepares the earth for the coming of our Lord and Savior, our Lord Jesus Christ.

Intercession for Traditional Churches

We need to pray for the traditional churches, that they will come back to their first love in Christ Jesus. We pray for a new breath of the Holy Spirit in these churches that have a great apostolic tradition. We pray for healing, that somewhere down the years, dry ritualism has replaced the lively, joyful power of the risen Christ and the freshness of the Holy Spirit in liturgy, bringing the glory and presence of the Lord back again.

Sometimes, the people in charge of the liturgy focus on the preciseness and the outward form, rather than its deep significance and the release of God's power at work through the liturgy. I have participated in many traditional churches in their worship and intercession, and have experienced the darkness, the coldness and dampness that so easily encroach on one's desire to pray and rejoice in God. I have watched people's facial expressions change from joy to one of being morose, when they entered the church.

One Saturday evening, August of 2010, I was at a Eucharist service in a traditional church. After we participated in Holy Communion, I had a vision of Elijah's chariot above the main altar. The entire roof of the church seemed to disappear before my eyes, and there was Elijah in his chariot of fire, above the main altar. I was so amazed at the vision, I kept staring at it and lost track of the liturgical service. I remembered the words of scripture from Malachi 4:6, that when Elijah returns at the end of the age, he will restore the hearts of the fathers to their children and the hearts of the children to the fathers. I began to intercede for this to happen in the church, for the city, and for the nation.

A week later, I attended another communion service at the same traditional church, and the first scripture reading was from the book of Malachi. It was about the return of Elijah at the end of the age. The Pastor's sermon was on the ministry of Elijah, who foreshadowed the Lord Jesus. He then asked the congregation a question, "What if Elijah visited our church today, what would he have us do?" I smiled to myself and thought, "He has visited your church just last week."

Through the reading of the scriptures at that service, I think Elijah wanted the church to pray for the reconciliation of families, within individual families, and with one another, especially fathers to their sons and sons to their fathers! The implications are more than just a family reconciliation. It speaks about the moral depravity of nations with millions of fatherless children, a sign of the end times and the proximity of the return of the Lord Jesus Christ. It was a call for that church to cry out in intercession for its city and its nation. Unfortunately, there was no place in that liturgical service where I could have shared this vision and called the people to intercession. At the end of the service, I spoke to the pastor about the vision. I shared with him what it meant for this church in and through its liturgy. I said the scriptures called us to intercede by the word of the God to cry out for our nation, and we missed it because we stuck to a liturgical format in word and not in spirit and truth. He replied, "Well, thank you for coming to our church, at least one man prayed according to that scripture reading."

When the church is bound to the outward performance of the liturgical service, the opportunity to move in the power of the Holy Spirit is lost, as it was in this church. The days are coming when the church must move in the power of the Holy Spirit. Dead traditions could lead to a dead church, for the Lord walks among the candle stands, and He will put out the sanctuary light of those churches who will not respond to His end-time call.

Heaven has stopped its worship to listen to the bride on earth respond to her Lord. There is silence in heaven to listen to the true "liturgy of love" from the heart, calling to her lover with real passion from the depths of her heart... "Come Lord Jesus, come!" Heaven waits in silence to hear the voice of the bride...then the Lamb opens the seventh and final seal. The Lion of the Tribe of Judah roars His final judgments to take back what is His.

The Seventh Seal is opened

With the opening of the seventh seal, the Lamb initiates the processes that destroy the enemy of God, and his followers. The Lord wins back the physical earth for Himself.

A Psalm of David: "The earth is the Lord's, and all it contains, the world, and those who dwell in it" (Psa. 24:1).

"*For the Lord Most High is to be feared, a great King over all the earth.*"(Psalm 47:2).

Summary of the Seventh Seal Judgments on the Earth

When compared with the Old Testament, the Exodus judgments on Egypt show us the Midrashic pattern of the book of Revelation.

TRUMPET Judgments in Revelation	BOWL (after the 7th trumpet is blown) Judgments in Revelation	THE PLAGUES ON EGYPT In the book of Exodus
First Trumpet Blown: Hail and fire burn up one third of the earth, one third of the trees, and all green grass burned (Rev.8:7)	**First Bowl Poured** on the people who had the mark of the beast and worshipped his image. They receive horrible, loathsome sores on their bodies (Rev.16:2).	Hail and fire burn up every plant and tree, yet Pharaoh does not repent. (Ex. 9:22-24).

Second Trumpet Blown One third of the sea becomes blood, one third of the sea creatures die, one third of sea ships sailing the seas are destroyed (Rev.8:8-9).	**Second Bowl Poured** on the sea. It becomes blood, killing all sea creatures (Rev.16:3).	The waters of the Nile become blood and the fish in the Nile died. Pharaoh does not repent (Exod. 7:17-21).
Third Trumpet Blown A star fell from heaven on one third of the rivers and springs. The waters become wormwood, killing those who drank the water, because it became bitter (Rev.8:10-11).	**Third Bowl Poured** on the rivers and springs, becoming blood (Rev.16:4).	The waters of the Nile become so foul, that no one could drink from it. Pharaoh does not repent (Exod. 7:17-21).
Fourth Trumpet Blown One third of sun, moon, and stars darkened (Rev. 8:12).	**Fourth Bowl Poured** on the sun, causing it to scorch men with fire. They blasphemed the name of God and do not repent (Rev.16:8-9).	Thick darkness covers Egypt for three days. No one could see each other. Pharaoh does not repent (Ex. 10:21-23).

Fifth Trumpet Blown Demonic warfare begins on the earth. Demonic locusts come forth from the bottomless pit, tormenting men only, not the vegetation (Rev. 9:1-12).	**Fifth Bowl Poured** on the throne of the "Beast," causes darkness to fall on his kingdom. They gnawed their tongues because of pain, as they blasphemed God and do not repent (Rev.16:10-11).	Locusts covered Egypt. Yet Pharaoh would not repent (Ex. 10: 14).
Sixth Trumpet Blown The demonic angels bound under the great Euphrates river are released to kill one third of mankind. The rest of mankind did not repent of idolatry (Rev. 9:13-21).	**Sixth Bowl Poured** on the Euphrates river, drying it up to make way for kings of the east. Frog-like demons gather the nations for the great war called Armageddon (Rev. 16:12-16).	Invasion of frogs from the river (Ex. 8:2-4).

Rev 11:15 Then the seventh angel sounded; and there were loud voices in heaven, saying, "The kingdoms of the world have become *the kingdom* of our Lord and of His Christ; and He will reign forever and ever. Amen"	**Rev 16:17 Then the seventh *angel* poured out his bowl** upon the air, and a loud voice came out of the temple from the throne, saying, "It is done."	Exodus 12:30: Pharaoh arose in the night, he and all his servants and all the Egyptians, and there was a great cry in Egypt, for there was no home where there was not someone dead. Exodus 12:31: Then He called for Moses and Aaron at night and said, "Rise up, get out from among my people, both you and the sons of Israel; and go, worship the Lord, as you have said.

In the Old Testament Exodus account, the Israelites who were delivered out of Egypt's demonic control become the betrothed of Yahweh, as He takes her to the Promised Land to dwell with her, just as a husband takes his wife to his dwelling place.

"*You will also be a crown of beauty in the hand of the Lord, and a royal diadem in the hand of your God. It will no longer be said to you, 'forsaken,' nor to your land will it any longer be said, 'desolate;'* ***but you will be called, 'My delight is in her,' and your land, 'married;' For the Lord delights in you, and to Him your land will be married. For as a young man marries a virgin, so your sons will marry you; and as the bridegroom rejoices over the bride, so your God will rejoice over you"*** (Isa. 62:3-5).

The "marriage of Yahweh and Israel" theme in Isaiah's prophecy is a foreshadowing of what Revelation presents us with in the "marriage of the Lamb" theme. Genesis and the book of Exodus offer a Midrashic pattern of the events that must take place after the seventh seal is broken in book of Revelation. In the book of Genesis, each of the six "day" creation events represents a certain specific order of creation. We know from Genesis 3, there are serious and catastrophic repercussions on the created order by the fall of Adam and Eve. In the book of Revelation, we find that God strips the earth of the created things in the very same order in which He created them. Only then can the new creation, the "re-Genesis" of the earth take place. The table below gives a bird's eye view.[48]

Genesis – Creation Account	**The Book of Revelation the 7th Trumpet**
The works of the six days of creation are:	Judgments over the created order
Day 1: Light (call it primordial light)	There is no judgment on the light. However, the in the "New Heaven" that comes down to the earth, (Rev.21:1, 23) the light from the glory of God dwells in the midst of man. It replaces the primordial created light of the sun.
Day 2: Firmament (or the heavens)	There is no judgment of the firmament. However, the new heavens and the new earth that comes down from heaven, replace the sky and outer space, as we now know it, with an expanse and starry host of a higher order. (See Rev 21:1, 23).
Day 3: Land and vegetation	First Trumpet Judgment on earth: Land and vegetation are affected.

[48] Read "A Rebirth of Images" by Austin Farrer, Wipf & Stock Publishers, 2006, for a much deeper exegesis.

Sea	Second Trumpet: Sea creatures are affected.
Water	Third Trumpet: Water is affected.
Day 4: The stars in the heavens	Fourth Trumpet: The stars and heavenly luminous objects fall from the heavens.
Day 5: Sea creatures and sea beasts	Fifth Trumpet: Sea creatures and beasts of the sea are affected.
Day 6: Land creatures and beasts	Sixth Trumpet: Land creatures and beasts from the land are affected.

*"Then the seventh angel poured out his bowl upon the air, and a loud voice came out of the temple from the throne, saying, '**It is done.**'"* (Rev. 16:17).

The Angel declares, **"It is done,"** as he pours the seventh bowl on the air. This induces the final process of the renewal and restoration of the earth.

*"Then He said to me, '**It is done**. I am the Alpha and the Omega, the beginning and the end. I will give to the one who thirsts from the spring of the water of life without cost'"* (Rev. 21:6).

Paul in his letter to the Ephesians calls satan the "prince of the power of the air" (Eph. 2:2). The angel pouring the contents of the seventh bowl on the air (Rev. 16:17), deals with the very source of the earth's sin and corruption. God will absolutely and completely, finish His work of purging all of the creation on the earth of sin and unrighteousness.

This complete purging of the atmosphere over the earth by the activities of the seventh angel makes way for the "New Jerusalem" to come down on the earth. The New Jerusalem is more than a city; it represents the tabernacle of God, the dwelling place, the "new" earth, that has no more sin. The book of Revelation symbolically represents the sea as sin. From out of the sea, the beast comes, and

Babylon is thrown into the sea. The book of Revelation in saying "there is no longer any sea," assures us that part of the restoration and renewal process of the earth is the destruction of the source of all sin, the root cause of sin. When sin and its source is completely destroyed, the new heavens, the new earth and the new Jerusalem, the bride comes down from heaven to reside with her Bridegroom, the Lamb, the eternal Son of God, the Lord Jesus.

"*Then I saw a new heaven and a new earth; for the first heaven and the first earth passed away,* ***and there is no longer any sea. And I saw the holy city, the new Jerusalem, coming down out of heaven from God, made ready as a bride adorned for her husband"*** (Rev. 21:1-2).

The earth is now ready to receive the Lord, and all those raptured with Him come back to the earth. The angel brings the bride of Christ to commence the marriage feast of the Lamb.

"*Then one of the seven angels who had the seven bowls full of the seven last plagues came and spoke with me, saying, 'Come here, I will show you the bride, the wife of the Lamb.' And He carried me away in the Spirit to a great and high mountain, and showed me the holy city, Jerusalem, coming down out of heaven from God, having the glory of God. Her brilliance was like a very costly stone, as a stone of crystal-clear jasper"* (Rev. 21:9-11).

Light and the firmament have no need of the created sun or moon in the new heavens and the new earth. The glory of God, walking with man again, is the perfect light, the Lamb of God, the Lord Jesus Christ. Jill, in her CD "Sounds of Heaven," captures the essence of the scripture so beautifully put into the lyrics in the song "You are Worthy." The lyrics read: "*Yeshua, You are fairer than the sons of men, Your beauty shines brighter than the sun in its splendor*".

"*I saw no temple in it, for the Lord God the Almighty and the Lamb are its temple. And the city has no need of the sun or of the moon to shine on it, for the glory of God has illumined it, and its lamp is the*

Lamb. The nations will walk by its light, and the kings of the earth will bring their glory into it. In the daytime (for there will be no night there) its gates will never be closed; and they will bring the glory and the honor of the nations into it; and nothing unclean, and no one who practices abomination and lying, shall ever come into it, but only those whose names are written in the Lamb's book of life." (Rev. 21:22-27).

The perfect will of God in heaven has come down on earth, and the Lord's prayer: "Thy kingdom come, Thy will be done on earth, as it is in heaven," is now fully consummated.

Intercession

"*Where there is no vision, the people are unrestrained, but happy is he who keeps the law*" (Pr. 29:18).

The church in the end-times must be a beacon of light in a world of frightening chaos. Her role will be like that of the prophet Elijah, a voice that declares the word of God, even in the midst of persecution. In order for her to carry out her task, she needs an end-time prophetic vision, and an understanding of the role that she must play before the return of the Lord Jesus to the earth.

"Behold, I am going to send you Elijah the prophet before the coming of the great and terrible day of the Lord" (Mal.4:5).

"And His disciples asked Him, 'Why then do the scribes say that Elijah must come first?'" (Matt.17:10)

The Prophetic Church acts Corporately

No one church or denomination will have the ability to move alone in the days ahead. There will be the need to work corporately. Those churches that continue to keep separate, and refuse to unify, may not be able to live through the dreadful times that lie ahead. In the humility of being part of the corporate body, will lie her power, grace and strength to carry out her ministry and to do the greater works of God (see John 14:12).

"Therefore if there is any encouragement in Christ, if there is any consolation of love, if there is any fellowship of the Spirit, if any affection and compassion, ***make my joy complete by being of the same mind, maintaining the same love, united in spirit, intent on one purpose****. Do nothing from selfishness or empty conceit, but with humility of mind regard one another as more important than yourselves; do not merely look out for your own personal interests, but also for the interests of others.* ***Have this attitude in yourselves which was also in Christ Jesus, who, although He existed in the form of God, did not regard equality with God a thing to be grasped, but emptied Himself, taking the form of a bond-servant, and being made in the likeness of men. Being found in appearance as a man, He humbled Himself by becoming obedient*** *to the point of death, even death on a cross. For this reason also, God highly exalted Him, and bestowed on Him the name which is above every name, so that at the name of Jesus every knee will bow, of those who are in heaven and on earth and under the earth, and that every tongue will confess that Jesus Christ is Lord, to the glory of God the Father." (Phil2:1-11).*

A corporate identity need not be a loss of individual identity. It is a way to work together in deep love, respect, honor and unity, with a common goal in mind, the mind of Christ. This holy unity expresses itself in the spirit of servant hood. When the church lives out this expression of servant hood, being one prophetic church in the city, God will grant her the awesome power to call down the power and judgment of God on the powers of darkness, which control the city. Pastors and church leaders must increasingly work out ways to express this unity as humble servants, not puffed up with pride about what each one has accomplished in "his church."

No church belongs to any pastor, or elder. The church belongs to God, to her Lord and Savior, the Lord Jesus Christ. Upper most in the mind of the pastor should be the desire to express the unity of the entire body of Christ in the city, and strive to achieve it at any cost. This expression of unity could work itself out in many different ways, but the most important way should be the expression of worship, warfare and intercession for the cities and the nations. Holding regular monthly meetings of all the churches in a city for worship and intercession, is the most powerful expression of God's awesome power at work, bringing down the strongholds in the city and setting captives free, healing the sick, opening blind eyes, and the lame to walk again. For in this place, the favor of God rests as the corporate church declares the word of God in all power.

Recently, I was in Vancouver, Canada where I attended a pastors' meeting. Pastors from **some** of the churches in the city of Langley met, shared, and prayed together for each other, their churches, and their city. I was witnessing what I sense is the end time vision of the church on a small scale. More pastors need to join or work together in a spirit of unity. The greatest reason for such unity should be more than just praying for "my church program." When pastors meet, their deepest level of purpose should be the desire to have a corporate expression of the church in the city, on a regular basis. The days ahead will need that support and wisdom to work together, in a city thrown into total chaos. This is the hour for the church to prepare for those times.

The Prophetic Church is God's "House of Prayer," not a "House of Performance"

It is time for the church on earth to be a powerhouse of prayer. The church will become a place from which the judgments of God will go forth against sin and unrighteous acts. This prayer will bring the powers of darkness down. It is time for Christians to make a decision as to why they go to church. The church leadership must make a decision as to why Christians should attend Sunday service. It would be most inappropriate to attend Sunday church service because the church has a great music ministry, or the pastor preaches great sermons on Sunday, or serves great coffee after the service, or for any other reason other than being God's powerhouse of prayer.

Through worship and intercession, the bride is making herself ready to reign and rule with the Lord Jesus in the coming millennium reign. It is time to be passionate individually as a Christian, and corporately as church, for the Lord Jesus' return to the earth. We need to come to that place in our hearts and minds where we come to church because we are His bride. As His bride, we enforce His kingdom rule in all the earth through worship, warfare and intercessory prayer. As we do this, He will return to the earth quickly, to dwell with us. Each time we meet corporately, His kingdom of righteousness, peace and joy in the Holy Spirit, will reign forever and ever in our cities and nations.

In the year 1992, while I was in India, communal riots between Hindus and Muslims broke out all over the nation. At that time, I lead a team of intercessors in conducting a monthly all-night intercessory prayer (1991-2001) for our cities, nation and church, in the heart of Mumbai city. The prayer vigil was a seven-hour period of worship, warfare and intercession, and it was always packed to capacity with earnest prayer warriors that came from as far as Kuwait and Dubai!

During the days of the communal riots of, the pastor of the church who hosted the night vigil of intercession called me and said, "Bob I think you will need to cancel the prayer vigil, as there is a police curfew in this area due to the communal riots." I replied that his suggestion was a good safeguard, but it would be hard for me to make telephone calls to the thousands that attended the prayer vigil. We would have to go ahead and trust God to bring all the intercessors safely, before curfew time, and to pray behind closed doors.

I was surprised to meet a police officer, who knew about our all-night vigils of prayer, and he called to ask me if I was going to pray for the riots to end quickly. I told him that was our intention. He said he would send a few more police officers to keep watch while we prayed!

That night the Lord told me to find two Christian converts from the warring factions of the riots, who would stand in the gap for their warring communities and repent. At the prayer vigil, I asked if there were any Christian converts from the two communities, who were suffering due to the communal riots. After a brief silence, two people came up to me. One of them had flown in from another nation where he was working. God had told him to go home on that weekend of the prayer vigil. The Lord told him to go home "to pray for a brief time, concerning the riot situation in his home town, Mumbai," as he had testified. I instructed them to "stand in the gap" as intercessors, and to ask forgiveness from the Lord for their communities, and to wash the feet of each other as a symbol of forgiveness. This would release of a spirit of love between the two warring communities.

The church resonated with deep sighs and sobbing, as intercessors prayed for healing of the heart of hatred in the hurting nation. Within less than a week, the riots all over the nation stopped completely and life returned to normal. I met the police officer who had called me prior to the prayer vigil, and I thanked him for his help. He remarked, "Please don't stop this prayer vigil, for it is doing so much good for the city and nation." How my heart wished that these words had come, not from a police officer, but from the pastors of the city. Will the church need to come to a point of persecution, before it learns the hard way for the need for corporate worship and intercession for our cities and nations?

When the church, in prophetic obedience, becomes God's house of prayer, it has power to defeat the powers of darkness that bind the city, and this sets the captives free.

The Prophetic Church Prepares for the Greatest Harvest

Every church needs to ask itself the question: "How do we get ready for the greatest harvest that is coming to our city?"

The Holy Spirit is waiting for the church elders/leadership to come before the Father and ask Him, "Father, show us how to prepare for the greatest harvest that You will send to your church, here in these end times." God will send revelation swiftly, because this is in His heart. The Lamb's delay in opening the seals that herald the end of the age is because the church has not made herself ready, neither for the marriage feast of the Lamb, nor for the coming harvest. It is time for every church to build their discipleship programs to love, care for and nourish the harvest, and to learn how to reign and rule with the Lord Jesus when He returns. The Lord desires that His churches in the city will work in corporate unity, sharing in the Lord's wisdom and resources, and working towards reaping the harvest of His children. It is important to build a support structure from a foundation that has a solid end-time perspective.

The Prophetic Church Joins Heaven in the Act of Restoration and Renewal of the Earth

"*And that He (the Father in heaven) may send Jesus, the Christ appointed for you, whom heaven must receive until the period of restoration of all things*" (Acts 3:20, 21).

There exists a hierarchy of evil, superimposed on societal structures. At the heart of this evil hierarchy is the destruction of God's laws, particularly pertaining to the family. From the sacrifice of children on the arms of Moloch in ancient times, to the "altars" of destruction in modern day abortion clinics, man has killed his own offspring, not sparing them even in their mothers' wombs.

I was awakened one night by the Lord, to pray for New York City. Through blurry eyes in the Spirit, I saw the word "NYC" hanging in front of me. I spent the next two hours praying in the Spirit for New York City. Later that day, the Lord showed me why. I read this headline on January 7, 2011: 'Chilling' statistics show 41% of New York City babies aborted.'

When Mother Teresa of Calcutta, India, visited the USA, the media told her she came to country that had no poverty and so she had no message for such an affluent nation. She told the media that the worst kind of poverty would be to take the life of a baby, so that we could live life as we wanted. The NYC statistics proved her so right.

Our city leaders, elders and leaders, have allowed the corporate houses of Mammon to destroy babies through abortion, and then harm our young children through junk food sales, and have allowed sexualized and addictive cultural images to destroy our teenagers with sex, alcohol, and demonic music. The lust for more money never ends, the means justifies the ends, and governments are happy to pass laws to benefit the greedy. They have passed laws that have destroyed marriage and family. God will now have His turn to destroy the destroyer and those who have followed him. The church is God's instrument to do this. Worship, warfare and intercession are God's ways in heaven, to bring His judgments down.

The church, like Israel, must stop wavering between the world's enticements and the Lord's tasks for the church at the end of this age. She must take her rightful place and call down the Lord's righteous judgments against the altars of modern day Baal in every city. They have built altars to the Babylonian gods of prosperity, as we see in the image of the "bull and bear" markets. We have lived complacently in a sinful demonic system, which is destroying family, church and nation.

It was not good for a nation with a Judeo-Christian heritage, when a Congressman stood before the U.S. House of Representatives, and prayed to the "angel in the whirlwind," for spiritual guidance for the U.S.A. (February 28, 2001). Additionally, one of America's recent presidents, in his inaugural address said, "When our Founders declared a new order of the ages, they were acting on an ancient hope that is meant to be fulfilled," and he referred to an angel that "rides the whirlwind and directs the storm?"[49] Even Washington's Monument is an Egyptian Obelisk. There are thousands of obelisks in the U.S., which are ancient Egyptian monuments to *"Ra,"* the sun god. Some even have hieroglyphics written upon them. In addition to America's precious Judeo-Christian heritage, there were also strong influences from occultist elements, who also came over from Europe, but with different motives than those of the Puritans.

Yes, there are altars to the ancient Egyptian and Mesopotamians gods, invoked by people in high offices, the "Pharaohs" of modern day governments. The seventh seal in the book of Revelation, when broken, will deal with these Pharaohs, and their plans and aspirations will meet with the wrath of God. The church at this last minute must arise to meet the task set before it by her Lord.

The church must have a priority of intercession at this end of the age. Just as the prophet Isaiah said, God still looks for a man. Even today, He looks for His bride to join Him in the task of the restora-

---

[49] Go to: http://www.raidersnewsupdate.com and find Thomas Horn's book, "Apollyon Rising 2012." I, the author of Worship, Warfare and Intercession before the Throne of God, do not necessarily agree with all of Mr. Horn's points or conclusions, but these quotes from U.S. public figures, were accurate.

tion and renewal of the earth. Will she join with heaven's intercession, so that He can bring His righteous judgments, so that He can bring, the kingdom of God on the earth?

CHAPTER 8

DEALING WITH GIANTS AND IDOLS

Hindrances to Entering the Worship, Warfare and Intercession of Heaven

As we move towards the coming of the "New Heaven and the New Earth," we will need to take a leaf out of Israel's history from the bible. There are important lessons to learn from them as they moved towards the Promised Land. Paul explains this fully in 1Corinthians 10. He warns the church that everything that happened to Israel was meant to serve as a warning and an example to the New Covenant church.

We live in the age of fulfillment of all of the Lord's plans. In the previous chapter, we learned the effects of opening the seven seals on the scroll in heaven. The seals release the judgments of God upon the earth. These judgments will bring forth the utter defeat of the enemies of God and the enemies of the true church. When judgment begins, and the process of stripping the earth takes place, it will be a time of great hardship and testing on the earth. For the church, it will also be a test of faith for all who follow the Lord Jesus.

"Each man's work will become evident; for the day will show it because it is to be revealed with fire, and the fire itself will test the quality of each man's work. If any man's work which he has built on it remains, he will receive a reward. If any man's work is burned up, he will suffer loss; but he himself will be saved, yet so as through fire" (1Cor.3:13-15).

However, God's hand of protection will be upon those who have made Jesus the Lord and Savior of their lives and who live according to His word. The pressure from the evil one, which we will experience through human agencies, such as the government, or businesses, will be intense, and could cause a Christian to stumble. The purpose of this chapter is to help us understand the nature of those evil forces, which will come against a Christian. If we have a grasp of how the enemy operates, then with God's grace and help, we can come through the battle as "overcomers." This will require wisdom, prayer with fellow Christians, and we will need the Lord's angels to help us. As we learned in previous chapters, the guiding principle is to study the biblical patterns and cycles found throughout the history of God's people. We can then learn and be warned of what will hinder us, and we will have wisdom as to how to deal with it in the future, in the mighty name of the Lord Jesus.

After crossing the Red Sea, the Israelites spent a generational time span in the Sinai desert. There, God had to teach them to be His people. He had to strip them of the behavior that they imbibed while living as slaves under a cruel Pharaoh, in the land that represented a life of sin. That paradigm of life would not equip them take the Promised Land. To enter the Promised Land, they had to ***contend with seven nations, whose people were "like giants."***

These "seven giants" that tried to hinder Israel while on her way to the Promised Land, will now try to hinder the church while on its march to the glorious future God has prepared for her.

The words of the Lord Jesus echo the same spiritual truth:

"Now when the unclean spirit goes out of a man, it passes through waterless places seeking rest, and does not find it. Then it says, 'I will return to my house from which I came'; and when it comes, it

finds it unoccupied, swept, and put in order. Then it goes and takes along with it ***seven other spirits more wicked than itself****, and they go in and live there; and the last state of that man becomes worse than the first.* ***That is the way it will also be with this evil generation.***" (Matt. 12:43-45).

Christians will experience these same seven spirits, who will try to prevent them from reaching the "new earth." It is therefore important to understand the nature of these seven evil spirits and their dealings with Christians. This gives us discernment for what we need to contend with, when we experience their demonic oppression.

"*When the Lord your God brings you into the land where you are entering to possess it, and clears away many nations before you, the* ***Hittites*** *and the* ***Girgashites*** *and the* ***Amorites*** *and the* ***Canaanites*** *and the* ***Perizzites*** *and the* ***Hivites*** *and the* ***Jebusites,*** *seven nations greater and stronger than you.*"(Deut. 7:1).

"***You shall not be afraid of them****; you shall well remember what the Lord your God did to Pharaoh and to all Egypt.*" (Deut. 7:18).

Paul confirmed this when he taught the scriptures on the Sabbath day in the synagogue at Pisidian Antioch:

"***When He had destroyed seven nations in the land of Canaan,*** *He distributed their land as an inheritance.*" (Acts 13:19).

There were seven nations living around the Promised Land, and they were giants in stature, as compared to the Israelites. This description teaches us the spiritual truth that there are "spiritual giants." In Ephesians 6:10-13, Paul talks about them as principalities and powers, rulers, forces of darkness. This is satan's hierarchal structure, from where he launches his attacks on the church.

The Hittites

The Hittites probably originated from the area beyond the Black Sea. The Hittites first occupied central Anatolia, which is modern day Turkey.

The name Hittite means "fear" or "dread." The Hittites were a people who instilled physical fear in the Israelites. The Hittites had an aggressive military mindset. They were a great power in the area before the Israelites entered the land of Canaan. Their empire lasted from 1750 B. C. until 1200 B. C., when Moses led the children of Israel towards the Promised Land. Their empire collapsed due to internal trouble. They were descended from Heth, a son of Canaan, who was a son of Ham, one of Noah's three sons. Esau grieved his parents by marrying two girls from outside the family. They were daughters of Heth, or Hittites.

Intercession

We need to pray for the church as the Lord begins to strip the earth and things get ugly. Stock markets will begin to crash, and the crash of financial and monetary markets will soon follow. The domino effect can cripple a nation's economy. Fear can cripple us from taking the right action. Fear is a "spiritual giant." On one hand, the Fear of the Lord is good, as it acts as a deterrent and a safety-switch, to protect us from doing the wrong thing, which could bring our lives into real danger. However, natural fear, or the fear of man, or of catastrophic events, prevents us from doing what God wants us to do in a particular situation.

I was once asked to be part of a prayer team to go to Rwanda, Africa. At that time, there was a war going on between two warring Christian factions, the Tutsis and the Hutus. Millions were being hacked to death. It did make "reasonable" sense not to go to the land when such a situation was prevalent. Yet, since it was a call from the Lord, I should have joined the team who went and prayed. I succumbed to the spiritual giants of fear, and I repented before the Lord for not going. The team was safe while they were there – they had some hardships, such as no food or water for some time. However, this was only a minor discomfort, because the teams where fasting, as part of this spiritual battle. They prayed for the nation and the war did stop.

The days ahead are going to demand a lot of resolve to be God's people, and to do the right thing in the eyes of God. No one in Noah's day went into a boat-building business in a desert land where it did not rain. It may not be what everyone is doing or following, but it is the thing God would have us do. One may have to pay the price of that testimony. God will give you the victory for your obedience, as you triumph through it. The Lord will crown you with the victor's crown.

One aspect of fear is compromise. Hardship, persecution, loss of lifestyle, loss of assets, loss of wealth, or other losses, can cause us to compromise our beliefs. This is the "Hittite giant" spirit at work. Supporting one another in times of hardship and praying for one another will help Christians overcome this evil spirit. Churches with household or home cell groups (smaller support groups) will do better than those that do not have such support structures. This is the time to build strong pastoral structure in the church, which is a way to help the flock to love and support each other in prayer, and with their resources. Church structures like these are better equipped to handle the coming harvest.

The other aspect of fear is oppression. Sometimes oppression can be a subtle, nagging thought or word spoken. It can come from an unexpected source, even from those who you thought were loving and kind towards you. There are times when we open doors in our life to the oppressor by our disobedience to God. Paul tells us that when our obedience is complete, we can make every oppressive thought subservient to the Lordship of Christ (1Cor.10:5).

In the days ahead, the oppressor will use every means in his evil arsenal to create paralyzing fear, to bring Christians to compromise. I have met people, who the moment they open their eyes, experience the fear of what the day has in store for them. Through a prayer ministry, deal with these fears as quickly as possible. It could entail deliverance of evil spirits, or healing of memories. Christians will need to build a determination that the Kingdom of God within them will prevail against the forces of darkness and they will not allow the oppressor to gain ground in their lives (Matt.11:12).

Fear can oppress us in a variety of ways. In nations where the laws of the country are against Christians, we can feel oppressed. In nations where Christians are in the minority, and the major religion is anti-Christian, fear can come from the places we work, secular or spiritual places. When people in authority, who work for unjust and personal gain, use fear to bend the will of the people under them, the Hittite spirit is in control. There is nothing sinful about making profit in business. However, greed often drives businesses to make profits at the cost of cheating, lying, and suppression. People working in such corporations must follow "company policy," which promotes unlawful means of gain. The Hittite spirit is at work in all these situations.

The same is true of laws passed by government, local or state, that may want to control the Christian faith by passing laws prohibiting freedom of religious expression to Christians, while allowing it to others. For example, in a New Zealand public school where my son attended, the authorities prohibited Christians from having a prayer meeting room, while Muslims in the school used a "prayer room" for their Friday prayers. This is just one simple example of this type of persecution prevalent in Western nations today. Most of these nations have strong laws against, "hate speech." However, one finds that these laws are used against Christians, rather than against those who practice hate speech, like the fundamental Islamists.

Churches must pray to overcome the spirit of compromise and resist the devil's oppression in the days ahead. We will need to pray for the Christians around the world, but especially in Israel, the Middle East, and in Islamic dominated nations. We may also find the need to pray for Christians in Europe. With increasing militant Islam, there will be a great polarization of beliefs that will lead to the persecution of Christians. Author George Otis, Jr. in his book "The Last of the Giants," (Chosen Books, Baker Book House, Grand Rapids, Michigan, 1991) writes about "The Reemergence of the Prince of Persia."

Here is a world news report:

London: The number of Britons choosing to become Muslims has nearly doubled in the past decade, (this is true of New Zealand too!) according to a study by an inter-faith think tank. The study by think tank Faith Matters attempts to estimate how many people have embraced Islam. Despite the "often negative" portrayal of Islam, thousands of Britons are adopting the religion every year, The Independent reported. [50]

Our God is greater than any giant, but the church must return to what the Lord Himself called her to be: a "house of prayer." Prayer is the only means by which Christ's victory on the earth is enforced. Paul said this to the church in Corinth, a rich and influential church in its day.

"*For though we walk in the flesh, we do not war according to the flesh, for the weapons of our warfare are not of the flesh, but divinely powerful for the destruction of fortresses. We are destroying speculations and every lofty thing raised up against the knowledge of God, and we are taking every thought captive to the obedience of Christ*" (2Cor. 10:3-5).

The Girgashites

In Genesis 10:16 and 1Chronicles 1:14, the tribe of Girgashites is listed as part of the mixed population of Canaan. The meaning of the word is "one who returns from a pilgrimage," or "dwellers in clay soil or black mud." The two meanings point to the same enemy, with the same intent of taking people back to where they were, before it all happened. There is an expression that says, "I'm stuck in the mud." The "Girgashite spirit" is the "turning back spirit" or the spirit of apostasy. The earliest introduction to the operation of this spirit is when Israel was delivered from Egypt. Egypt was a harsh land, and it was here that the Israelites were enslaved by a cruel Pharaoh. The Hebrews knew only slavery and the lash. Canaan was the Promised

[50] Read the full article at: http://articles.timesofindia.indiatimes.com/2011-01-04/uk/28352947_1_converts-conversions-batool-al-toma

Land, flowing with milk and honey. It was also the place of freedom for the Hebrews. They were in a land free from the oppression of the past. It was also freedom from the oppression of the demonic gods of Egypt. It was a place to worship the one true God.

This freedom, both physical and spiritual, was a foreshadowing of the "new earth and the "new heavens" which the church will experience when delivered from the "pharaohs" of the powers that control this world system. Astonishingly, the Hebrews wanted to turn back to Egypt. Influencing this decision, were the rebellious leaders whom the people chose (see Nu.14:4). Behind the desire to turn back lies the sinister hand of the "Girgashite spirit." The people foolishly said, "*Would that we had died in Egypt where we had plenty to eat.*" (Exodus 16:3).

Once the Israelites crossed over into the land, after the unbelieving generation had died off in the wilderness, the Girgashite spirit was heard of no more.

The Spirit of Apostasy Turns a Christian Back from Following the Lord Jesus

When Christians do not make the Lord Jesus the center of their life, everything else will work to get that central place in their heart.

"*Do not love the world nor the things in the world. If anyone loves the world, the love of the Father is not in him. For all that is in the world, the lust of the flesh and the lust of the eyes and the boastful pride of life, is not from the Father, but is from the world.*" (1John 2:15-16).

The Girgashite spirit seeks to turn a person from his inheritance in Christ. Jesus said that no man can serve two masters. This spirit will try its best to influence us to mistrust God in the days ahead, when the shaking and stripping hit us hard. Just as the Israelites desired to leave the desert to go back to Egypt, Christians also might long for the "good old days." When we look back, God has no place in our lives; we are just offering lip service.

What is your strategy for these days? Are we living for our own pleasure and entertainment? Jesus spoke about the "days of Noah," which was an accurate description of how people lived for enjoyment and personal pleasure, while their spiritual condition was rotting. In Genesis 13: 10-11, Abraham had given Lot the choice of the land, when they decided to separate to their own ways. Lot looked down into the fertile Jordan Valley and realized that he could make more profit there than in the hills of Judea. There was city life for his wife and daughters, rather than the rural life of the hill country. When God decided to destroy Sodom, He sent angels to make sure that Lot and his family fled from Sodom before the act of destruction. As the family fled, Lot's wife looked back with a desire for the life she was leaving, and it led to her death. **In Luke 17: 32, Jesus warns: *"Remember Lot's wife."*** That is the price of apostasy: "*You are the salt of the earth; but* ***if the salt has become tasteless,*** *how can it be made salty again?* ***It is no longer good for anything, except to be thrown out and trampled underfoot by men"***. (Matt. 5:13).

Intercession

The Lukewarm Spirit in the Church

Love for the world is a spirit that takes a Christian away from the love of the Lord. When a Christian begins to love worldly pleasures, the Girgashite spirit is at work, by developing a lukewarm spirit to the things of God within the believer. The Lord Jesus spoke about in the parable of the prodigal son, who ran after pleasures and regretted it later.

Why does the Lord Jesus have strong words for those who are lukewarm towards God? It is because the lukewarm spirit is manipulative, and it appeases the world on one hand, and the Lord, on the other hand. A divided heart mocks God. **There are Christians who want the best of both worlds, the best of God's kingdom, and the best of the kingdom of darkness.** The Lord says that you can only serve one master, and you have to make a choice, as to whom you will serve. He will deal with a lukewarm spirit, by letting the world that you loved, deal with your destruction. It is good to heed the words of the Lord Jesus: *"Remember Lot's wife."*

"*I know your deeds, that you are neither cold nor hot; I wish that you were cold or hot. So because you are lukewarm, and neither hot nor cold, I will spit you out of My mouth*" (Rev.3:15-16).

The Christian must decide his allegiance. In a church with a strong pastoral structure, through prayer and support of a house group, a Christian can receive help to overcome the spirit of the world and all its empty enticements, which are passing away (see 1John 2:17).

Pray for the Nation

Praying for people in government and other places of authority is necessary and important. The Girgashite spirit works in governmental structures in a variety of ways to control the nation. However, its focus is in the law-making bodies of government. This evil spirit's main goal is to cause the nation to go against the Lord's laws. It is a spirit of "lawlessness". Paul says this spirit will prevail in the last days. (See 2Tim.3:1-5). Most nations have laws that have legalized abortion, "mercy killing" or euthanasia, and prostitution. These laws ensnare a person into wrongdoing. When Christians are ensnared in any of these activities, the danger of falling away from their faith in the Lord Jesus is real. Repentance, confession, and the love of the church can restore a fallen Christian.

The words of U.S. Supreme Court Justice Felix Frankfurter (1952) are a real revelation. He said, "The real rulers in Washington are invisible and exercise power from behind the scenes." Dr. Thomas Woodrow Wilson, the 28th President of the US (1913-1921), said, "Some of the biggest men in the United States, in the field of commerce and manufacture, are afraid of somebody, are afraid of something. They know that there is a power somewhere so organized, so subtle, so watchful, so interlocked, so complete, so pervasive, that they had better not speak above their breath when they speak in condemnation of it... (we are) no longer a government by free opinion, no longer a government by conviction and the vote of the majority, but a government by the opinion and the duress of small groups of dominant men."

It is not strange that Paul mentions in his "First of all" priority list to Timothy should be prayer for those in authority (1 Tim 2:1-8).

Pray for the Church

From Abraham's intercessory struggle to save Sodom and Gomorrah, we see a difficult lesson, which the church must learn. Our cities may suffer judgment, because the intercession rising up from the church for our cities is not enough. Did Christchurch and Japan suffer judgment, because intercession by the church in Christchurch and in Japan was not enough? We may not be able to answer such questions immediately, however, we must search our hearts to see if we, personally, have given priority to pray for the cities and nation, or if we have squandered that time on non-essentials. We will have to give an account of our stewardship of time.

Pastors have a responsibility before God, with regard to their ministry of prayer. They must make worship, warfare and intercession a priority over everything else. We are living in the end of ages. The four horsemen of the apocalypse are visiting the cities of the nations and have a mandate from God, to carry out His end-time protocol. The urgency to prayer is like the need for air to breathe. Pastors, like the elders in the heavenly worship that come with bowls of incense, the intercessory prayers of the church, should lead the intercession for their nation and cities and churches. Pastors should be the prime movers, calling the flock to intercession. The time has long past, when intercession was a back room ministry of retired folks with time on their hands!

There was a time when the pastor would make the announcement: "If any of you have time on Tuesday evenings, a small group meets in the back room to pray for the city and the nation for about thirty minutes, and after that, there is great fellowship with coffee and cookies for the next hour, before we close up."

Will we achieve the required numbers of intercessors? Abraham could not find ten men that would pray in his day. Even as a friend of God, he could not save the fate of Sodom and Gomorrah. Many of our cities today, are a lot worse in their spiritual condition than those two cities. If God cannot find enough intercession going up, our fate will be similar to those of Sodom and Gomorrah. God is merciful and gives His church the opportunity to seek after His own heart for the cities and the nation. However, in His end-time timetable, the days are falling short, and only fervent prayer by the church can save the city from the coming wrath of God. Pastors, do not neglect prayer for your cities and nation.

The fate of Jerusalem rested on the intercessors for that city

"*As he approached Jerusalem and saw the city, he wept over it and said, If you, even you, had only known on this day what would bring you peace—but now it is hidden from your eyes. The days will come upon you when your enemies will build an embankment against you and encircle you and hem you in on every side. They will dash you to the ground, you and the children within your walls. They will not leave one stone on another, because you did not recognize the time of God's coming to you.* ***Then he entered the temple area*** *and began driving out those who were selling. It is written, he said to them, My house will be a house of prayer; but you have made it a den of robbers.*" (Lk 19:41-46 NIV).

- Jesus sees in the spirit the impending judgment of the city of Jerusalem
- He rushes off to the temple to find if there is prayer going up in repentance for the city, to look for a man that will stand in the gap for the city.
- But instead of prayer, there is commercialism in the house of God
- He is angered that God's house of prayer is turned into a "den of robbers"

In AD 70, the invading Roman army reduced Jerusalem and the Temple to rubble. There is a lesson for the churches today from this scripture and its fulfilment in history.

Looking for and Hastening the Processes that Lead to the Return of the Lord

For most of us, our theology has brought us to a place to believe that "God is good all the time." This is perfectly true and right. However, we must understand the terrible and destructive effects of humankind's sin upon the earth. Paul tells us that "*all creation groans, waiting for the revealing of the sons of God*" (see Romans 8:19). Isaiah tells us that "*the earth reels and staggers like a drunkard, so heavy is the weight of our sin upon the earth*" (see Isa. 24:)

Just as there are natural laws, which are unavoidable, there are also spiritual laws, which are fixed. If you jump off a cliff, you will fall to your death. This is not God's fault. You have violated a natural law. In the same way, if man has violated God's covenants and spiritual laws there MUST be consequences. God does not wish that any would perish or suffer this way. His Father's heart yearns to bless His own creation. Nevertheless, sin and death have entered the earth, because we, like sheep, have all gone astray. These last days' calamities are part of the healing and restoration of the earth, but they will be terrible to endure. We should not blame God for allowing these catastrophes, because even in the midst of them, He is merciful.

The Lord Jesus warned us in love, that earthquakes, tsunamis and the stars falling from the sky to the earth would occur in the last days. These are literal events, and the current trends are pointing in this direction. The Lord told us that when we see all these things happening, *"Lift up your head, for your redemption is near."*

"There will be signs in sun and moon and stars, and on the earth dismay among nations, in perplexity at the roaring of the sea and the waves, men fainting from fear and the expectation of the things which are coming upon the world; for the powers of the heavens will be shaken. "Then they will see the Son of Man coming in a cloud with power and great glory. But when these things begin to take place, straighten up and ***lift up your heads, because your redemption is drawing near*****"** (Luke 21:25-28).

The book of Revelation outlines the processes that God will use to bring changes to the earth and the heavens before He comes again. The Lord is beginning to execute His righteous judgments, which bring great and fearful shakings upon the earth, the prophets often spoke about the terrible day of the Lord.

"*Behold, it is coming and it shall be done," declares the Lord GOD. "That is the day of which I have spoken."* (Ezek 39:8).

"The LORD utters His voice before His army; Surely His camp is very great, For strong is he who carries out His word. The day of the LORD is indeed great and very awesome, And who can endure it?" (Joel 2:11).

He desires and expects His bride to agree with Him in His righteous judgments, through worship, warfare and intercession. What pulls us back from aligning ourselves with the Lord in this process, is the "Girgashite spirit." This spirit pulls us away from the "Promised Land."

The Amorites

The Amorites were members of an ancient Semitic-speaking people, who dominated the history of Mesopotamia, Syria, and Palestine from about 2000 to about 1600 B.C. The Amorites were a large and powerful nation that controlled much of the Promised Land, including Jerusalem, Hebron, and Lachish (see Joshua 10:5).

The word "Amorite" means to speak against, or to boast of oneself. The name Amorite comes from the Hebrew name "Amor," which is identical to the verb "*amar*" meaning: to say, speak, intend, to promise.[51]

In a study from the rabbinical and apocryphal literature, the Amorites are described as a people of witchcraft: A special section of the Talmud[52] is devoted to the various superstitions, called

[51] For a deeper understanding refer http://www.abarim-publications.com/Meaning/Amorite.html

[52] Tosef., Shab. 6-7. [7-8.]; Bab. Shab. 67*a et seq*.

"The Ways of the **Amorites**." The Book of Jubilees[53] says that, "the former terrible giants, the Rephaim, gave way to the **Amorites**, an evil and sinful people whose wickedness surpasses that of any other, and whose life will be cut short on earth." In the Syriac Apocalypse of Baruch[54] they are symbolized by "black water," on account of "their black art, their witchcraft and impure mysteries, by which they contaminated Israel in the time of the Judges." [55]

The two opinions together give a clear picture of the "spirit of the Amorites." It is an evil spirit that uses speech or words in witchcraft (a common modern word used today is "*mantra*") to cast spells and to work evil in the lives of others, which is one form of sorcery. Many people have at some time, been bothered by the speech of another. Gossip, slander, and careless and thoughtless words are used, with intent to harm a person's name, character, or wellbeing. One way in which this spirit operates is through covetousness and seduction.

Sorcery and witchcraft are common practice even today. In the upper echelons of corporate power and in political circles, sorcery and witchcraft are used to gain places of wealth, position and power. The word of God condemns sorcery and witchcraft:

"*But for the cowardly and unbelieving and abominable and murderers and immoral persons and* ***sorcerers*** *and idolaters and all liars, their part will be in the lake that burns with fire and brimstone, which is the second death.*" (Rev. 21:8).

So powerful is the faculty of speech, that Solomon declared: "*Death and life are in the power of the tongue, and those who love*

[53] The Book of Jubilees (chapter 29: [9] 11), is an extra biblical, almost extinct writing from the "Pseudopigrapha."

[54] The Syriac Apocalypse of Baruch (chapter 60), is an ancient Jewish text from the Pseudopigrapha, written in the late first century after the destruction of Jerusalem in 70A.D.

[55] Read more http://www.jewishencyclopedia.com/view.jsp?artid=1422&letter=A&search=amorites

it will eat its fruit." (Proverbs 18: 21). The Living Bible puts it this way: "*Men have died for saying the wrong thing.*"

The Lord Jesus was addressing the Amorite spirit, which was attacking the people of His day, when He said, "*For by your words, you shall be justified and by your words, you shall be condemned*" (see Matt. 12:36-37). Paul, in his epistle to the church in Ephesus, says: "*Let no evil talk come out of your mouth, but only such as is good for edifying, as fits the occasion, that it may impart grace to those who hear*" (Eph. 4:29).

James gives us an even stronger emphasis:

"*So also the tongue is a small part of the body, and yet it boasts of great things. See how great a forest is set aflame by such a small fire! And the tongue is a fire, the very world of iniquity; the tongue is set among our members as that which defiles the entire body, and sets on fire the course of our life, and is set on fire by hell. For every species of beasts and birds, of reptiles and creatures of the sea, is tamed and has been tamed by the human race. But no one can tame the tongue; it is a restless evil and full of deadly poison. With it we bless our Lord and Father, and with it we curse men, who have been made in the likeness of God; from the same mouth come both blessing and cursing. My brethren, these things ought not to be this way.*" (Jas. 3:5-10).

Intercession

How can we intercede for our cities and nation, where hundreds are bowing down to the Amorites spirit? From politicians to pagans, and even Christians, all are guilty of using the tongue with bad intent. Christians must repent if they have done so, and pray a blessing instead. Intercession for the cities and nations must include dealing with curses.

- Ask the Lord to send angels that will deal with this spirit at every level of society.
- Bind curses made over the city:

    - By national and state leaders, or foreign dignitaries who may have placed a curse over it.
    - By business people who seek to gain something, from cursing the city or the nation.
    - By those who deal in the occult powers (witchcraft and all its related expressions like crystal ball gazing, etc.)
- Pray scriptural blessings over the city and the nation. Make a list of those who are in government, and pray a prayer of protection over them and their families. Pray that God would put a restraint over them, that they would not engage in sorcery and witchcraft, in order to gain an unfair advantage, to earn fortune and fame.

The Canaanites

The Canaanites, with whom the Israelites came into contact during Joshua's conquest, were a sophisticated agricultural and urban people. The word Canaanite means "trafficker" or "trader." The Canaanites were well established in the land for a long time as a trading people. They had been there since 2000 B. C., before the days of Abraham. In fact, Abraham came to live amongst them, when God called him out of Ur of the Chaldeans to travel to the west.

The perversion that came into the exclusive Israelite religion from the Canaanites, included distorted ideas of how their gods brought renewal and restoration to the earth. Canaanite myths incorporated fertility myths, found in mythological texts of the ancient city of Ugarit (modern Ras Shamra) in northern Syria. In this corrupt mythological system, Baal and his impure sister-consort are significant in the creation of the world and the renewal of nature, through abominable acts involving temple prostitutes.[56] Satan twists the biblical truth of the restoration of the earth, through detestable, unclean myths found in many pagan religions. These include orgies as part of their rituals. The Canaanite spirit is a spirit of perversion of God's divine laws and truths.

---

[56] For more research on this topic see http://history-world.org/canaanite_culture_and_religion.htm

Much of modern day advertising has a sexual, immoral spirit behind it. It allures young minds to lust and desire. The Canaanite spirit is the "trafficking spirit," the spirit of greed and lust for material goods, encouraging man to love mammon. Mammon was a Canaanite god of prosperity and wealth. The Canaanites worshipped Mammon and thanked him for their prosperity, instead of giving thanks to the true and living God. The Canaanite spirit is at work, enticing people to give their loyalty to money and power, rather than to God. The Canaanite spirit promotes a lack of contentment and a striving for more and more. People always seek a better computer, a better mobile phone, a better car, a better home, and the "better" list seems endless. Business is worshipped, and therefore, all kinds of deceptions to increase profits are endorsed. Lies, trickery and ruthlessness are all recognized practices in business. Morality is sacrificed on the altar of the god of business.

God has struck the house of Mammon and it has shaken. The financial crisis of 2009 revealed the corrupted network of the rich. Few nations have escaped the ripples of the financial crisis. God will continue striking the house of Mammon, until it completely falls. Paul says in 1Tim. 6:10, *"For the love of money is the root of all evil."* The Lord Jesus said that we must choose between Mammon and God (see Matt.15:21–28). In other words, the love money is the defining issue in measuring our devotion to the Lord. Some years ago, I served as a pastoral elder of a covenant Christian community. There were times when the most difficult area to disciple was the subtle love for money in the believers' lives. I have seen Christian families split over the issue of property unfairly inherited or unfairly distributed. He will shake the house of Mammon until it comes tumbling down, and with it will be many nations, ungodly businesses, and even the lives of those who built their fortunes from the house of Mammon.

Intercession

Scriptures tell us that these are principalities and powers, rulers of darkness in high places. Nations are under their control. God, when dealing with the house of Mammon, sometimes has to destroy the infrastructure that supports the work of Mammon. The implication of this is so huge, that it may be hard to comprehend, and we may ask, could not God do it another way? The answer is NO. Only God knows how to deal with these rulers of darkness over nations, and the way He chooses to do it is indeed the best way.

For most of the millions under the oppression of Mammon, we need to pray with a compassionate heart. This spirit of false religious beliefs blinds many nations. Power and wealth, false ideologies, personal glory and pomp, are some of the ways this spirit binds people. Many people blinded by the spirit of Mammon, are caught in the vicious circle of sex, drugs, debt, and gambling. Entire families end up destroyed by the debt trap. Legalizing gambling appears to bring in revenue, but it actually bankrupts families and robs society of economic health. **The advertisements glamorize gambling, but under the glitz and glitter, lies withering poverty.**

The nations will experience God's shaking and stripping of the perverted systems of man, which have so enslaved millions, unto desperation and ruin. God has promised that He will wipe away their tears and give them a new opportunity. He did this for the people of East Germany, of the U.S.S.R., and many other nations. It is foolishness and hypocrisy for governments to endorse, license and advertise various forms of gambling, when families are destroyed by the wickedness of the system, then they offer counseling for the addictions. This is the wisdom of the foolish. God in His wrath will destroy this wicked system. It must fall, so that the powers behind it are exposed. Then God will restore and renew the earth.

A time to weep for our nation, cities and the church. The Lord Jesus wept over Jerusalem. We need to pray that God will stay his hand a little longer, so that the church gathers the harvest. We are moving into a time of chaos. As nations declare bankruptcy, citizens will succumb to the chaos. We have seen riots break out in several nations in Europe in 2010. Italy, Paris, the U.K., Greece, to name a few; have had riots, due to financial crisis. Thousands are unemployed, due to a falling economy, and selfish business policies fuel the fire of civil unrest. What we are beginning to see in these days is a recipe for the disintegration of society, which has occurred in every previous age. These events signaled the downfall of empires and the closing of an age. The book of Revelation prophesies this pattern, as the age of the kingdom of God dawns.

The Perizzites

The name "Perrizite" has two meanings.

Firstly, it means an unwalled town or village. The Perrizites were hill dwellers, implying that they were unprotected by walls. In biblical times, the city built walls around to protect it from the enemy. However, the Perrizites towns were an exception, built on a hill, they had no walls.. A city without walls was a city that an enemy could enter at will. It is a vivid picture of a spiritual reality.

A city without spiritual "watchmen on the walls," is a city that is open to the powers of darkness. The church is the only agent that can protect the city from the spiritual forces of darkness that can enter to destroy, and to take its inhabitants captive. The church alone is equipped to raise prophets and intercessors, pastors and elders, to defeat the spiritual forces of darkness. Many "passive" religions that offer peace, prosperity and wellbeing, have a large following, because rather than fighting the enemy, they feed it. Sometimes, their members, unfortunately, include Christians. These religions actually open the city to the influence of the evil forces.[57] If the cities are places where there is increasing murder, violence, sex and drug abuse, they are unsafe. They are like the cities of the Perizzites, who have no spiritual walls to protect them from the spiritual forces of darkness. A praying and interceding church is a great challenge to the "Perizzites spirit." This spirit is restricted in its nefarious activities in the city.

The Church Must Pray for the City

"*And He came out and proceeded as was His custom to the Mount of Olives; and the disciples also followed Him. When He arrived at the place, He said to them, 'Pray that you may not enter into temptation.' When He rose from prayer, He came to the disciples and found them sleeping from sorrow, and said to them,* ***'Why are you***

[57] Read George Otis Jr. *The Twilight Labyrinth.* Published by Chosen Books, 1997

sleeping? Get up and pray *that you may not enter into temptation'"* (Luke 22:39-40, 45-46).

Many Christian groups and individuals in New Zealand, pray against the spirit of abortion in the nation. In 2010, the statistics indicated that the abortion figures went down. I am sure that all over the world, concerned Christians in every nation who have prayed against some form of evil, have seen the fruit of that spiritual warfare. What does it prove? Where there is fervent prayer by the church, it destroys the enemy's work and frustrates his plans.

If the churches in the city would come together and would pray regularly for the city, we would find safer cities. However, when it comes to churches unified together as "watchmen on the walls" for the city, the Lord Jesus finds the church "sleeping." He has a simple command to us: *"Do not be tempted, get up and pray."*

A second meaning of the name "Perrizite" is to Separate

It seeks to separate man from God. This is the deeper reason why this spirit will come against any church that will try to build a wall or hedge of protection, or to pray for unity among the churches. It knows that if the church attains unity, the city will turn to Christ and experience redemption. As long as the city churches remain separated by not praying together, it will keep the unbelievers from God, and will separate the believers from each other.

It separates one Christian church from another. In Judges 3:5-6, we see that Israel did not wipe out these races, as God had commanded them to do, but in fact, there was intermarriage. The affects of this disobedience played right through the history of Israel, down to this day. They have come against Israel and challenged her very existence. There is an important spiritual truth in Nehemiah's account of rebuilding the broken walls around the city of Jerusalem and the opposition from within.

David captured Jerusalem from the Jebusites. Much later, when the Israelites were taken into captivity in Babylon, the city had fallen into enemy hands. The walls had been broken down and the city of Jerusalem defenseless without its walls.

Nehemiah heard about the walls being broken down, and he sat down and wept. He repented and fasted for the nation of Israel for several days (see Neh. 1:4). He was the king's cupbearer, and he asked the king for permission to return to Jerusalem, to rebuild the walls of the city. His request was granted, and he set about his task. But the Perizzite spirit was at work, seeking to keep Jerusalem's walls from being built, so that it could keep the city of Jerusalem and its people, spiritually down trodden.

Sanballat and Tobiah went to great lengths to frustrate the plans of Nehemiah, who sought to rebuild the walls and to have Jerusalem useful for God's purposes again. Nehemiah 4:15-23 gives a very clear picture of the spiritual battle going on, while the physical work was carried out. There was certain physical work to do, but each person involved in the building had to be prepared for spiritual warfare. Why did Sanballat and Tobiah oppose the work going ahead?

The Perizzite spirit dominated the minds of the Israelites, through intermarriage with these nations, which were not completely destroyed in the conquest of the Promised Land. The "holy seed" mixed with the people of those lands. The leaders and overseers had been chief in this sin. Ezra 9:1-2 tells us: "*Now when these things had been completed, the princes approached me, saying, 'The people of Israel and the priests and the Levites have not separated themselves from the peoples of the lands, according to their abominations, those of the Canaanites, the Hittites, the Perizzites, the Jebusites, the Ammonites, the Moabites, the Egyptians and the Amorites.**For they have taken some of their daughters as wives for themselves and for their sons, so that the holy race has intermingled with the peoples of the lands; indeed, the hands of the princes and the rulers have been foremost in this unfaithfulness.***"

When Christianity marries the "things of the Spirit of God" with worldly wisdom, what we get is syncretism of theology, ministry, and worship. Syncretism means the mixing and mingling of religious beliefs, which are not compatible with each other. The spirit of the world enters the church. This "Perizzite spirit" will aim to keep the church divided. The result of that division in the church is a city with spiritual walls broken and open to the evil one.

What are the things of the Spirit of God, which have been married to the worldly spirit? What are the worldly idols, stolen and hidden by the people and leaders of the church? I believe that if we can be honest with these questions, we should see Christian unity rebuilt in the churches in the city. As united churches pray for the city, its spiritual walls will be rebuilt, and the awesome power of God will be displayed, as the enemy is cast down by the church by prayer and fasting.

Time Magazine in its September, 1990 issue, declared New York City to be the murder capital of America. In 1995, a movement of a prayer watch, called the "Lord's Watch" for the city, began among 45 churches in New York City. Today they have a website called Concerts of Prayer, which states that their vision and mission is to see the churches of New York and greater New York united in concerted prayer for the advancement of Christ's kingdom. Their mission is to mobilize pastors and churches to experience the awesome power of united prayer. One of the most spectacular results of this concerted, united praying by churches in NYC, was the dramatic drop in violence and crime: a 60% drop! N.Y.C. was then declared as the safest large city in the U.S.A. This is certainly a testimony to what can happen when churches in a city begin to pray together for their city and nation.[58]

Intercession

For Church unity in prayer

The judgment of nations begins with the judgment of the household of God: "*For it is time for judgment to begin with the household of God; and if it begins with us first, what will be the outcome for those who do not obey the gospel of God*?" (1Pet. 4:17). Christian unity seems to be something that only the Lord will achieve. Every church wants it, every leader talks about it, but few will labor to achieve it.

Can we say that the Perizzite spirit has outwitted the church, and kept her divided in fragments? We must concede to defeat, if we are

[58] Read "*The Power of a City at Prayer. What happens when Churches unite for Prayer*" McKenzie Pier and Catherine Sweeting, InterVarsity Press, 2002

to rise up and allow the Holy Spirit to work His unity amongst us. In some cities, there may be a lot more unity among the churches than in other cities. We can praise God for the pastors who have achieved some degree of unity. Unfortunately, the goal to intercede for the cities and nations, does not register in the agenda of many pastors. Can we hear the voice of the Lord Jesus saying, "*Father, the time has come*" (John 17:1).

The Hivites

The word Hivite means "life-giving through deception." Joshua 9 makes reference to the Hivites, who are called the Gibeonites in Joshua 9 :7. Israel made a treaty with them, without seeking the Lord. The Hivites deceived the Israelites into making a covenant with them, so that they would not be driven out of the land, or be killed.

The Hivite spirit is a spirit of deception. Biblical historians generally agree that all of the early descendants of Ham indulged in the most vulgar systems of worship. Historians believe that there was a deliberate departure from the true knowledge of God, which had been handed down from their father, Noah, who was the heir of righteousness (Hebrews 11:7). They chose to indulge in the iniquity of a debased and degraded society around them.

- The spirit behind the Hivites portrays this world with its illusions, delusions and dubious delights, as a life-giving, soul-satisfying environment.
- The Hivite spirit's objective is to ensnare people by cunning claims of providing pleasure and prosperity, at the price of their souls, rather than seeking pursuits that are of eternal value and worth.

In Matthew 16:26, the Lord Jesus warned against the spirit of the Hivites when He said,

"*For what will it profit a man if he gains the whole world and forfeits his soul? Or what will a man give in exchange for his soul?*

For the Son of Man is going to come in the glory of His Father with His angels, and will then repay every man according to his deeds" (Matthew 16:26-27).

The Lord Jesus Warned His Church of this Spirit in the Last Days

"Many false prophets will arise and will mislead many. Because lawlessness is increased, most people's love (of God) will grow cold" (Matt. 24:12).

The Apostle Paul Warned the Church of this spirit in the Last Days

"But realize this, that in the last days difficult times will come. For men will be lovers of self, lovers of money, boastful, arrogant, revilers, disobedient to parents, ungrateful, unholy, unloving, irreconcilable, malicious gossips, without self-control, brutal, haters of good, treacherous, reckless, conceited, lovers of pleasure rather than lovers of God, holding to a form of godliness, although they have denied its power"(2Tim 3:1-5).

Many are turning to all kinds of activities to give them a "high," or thrill, or to make it "big" in life. People yearn for the symbols of the high society, such as a membership to a prestigious golf club, or a house in the "million dollar plus" locality, or to drive a car that costs a hundred thousand dollars, or more. Sex, alcohol, wasteful spending, and the "big life" is projected as the superior lifestyle. It is largely the media, which creates and promotes this false picture and insatiable desires. Paul so rightly described the spirit of this age as "*reckless, conceited, lovers of pleasure rather than lovers of God.*" We have seen this reality each time an age and a debauched empire comes to its sorry end. Be it the Greeks or the Romans, they would rather find themselves dead in wanton debauchery when the end came, than in living holy lives. The ancient ruined city of Pompeii near modern day Naples, Italy, is a history lesson in solidified lava. The people of this city soaked body and soul into drunken debauchery, when the volcano Mount Vesuvius, erupted. The city was buried under 5 to 6 meters of hot, molten lava. The Lord Jesus remarked that as it was in the days of Noah, which was a generation of party revelers, so too is this generation, which will see the Lord's return. The Hivite spirit leads people to debauchery, drunkenness and destruction.

Proverbs 14:12: "There is a way, which seems right to man but the end is the way of death."

Intercession

The Hivite spirit is a spirit of deception. Deception is seen at all levels of society, in every nook and corner of life. This is a manipulative spirit, where the end justifies the means. We see this deception in many infrastructures of society: science, law and order, education, business, or charity – these institutions can subtly misrepresent facts, figures, data or findings, for their own benefit. This does not mean that every educational institution or every charitable organization misrepresents statistics. However, it does mean that within the entire spectrum of society, the "Hivite spirit" roams, looking to deceive and to pull us away from objective truth.

Truth sets us free; deception binds us to lies, dishonesty, and dubious ways of life. The intercessor needs discernment, as we sort through the different structures that make up modern society. As you pray, you will find this spirit hiding. As you discern it, ask the Lord to handle this spirit. Send it to the throne of Jesus for His judgment. Set the captives free. A note of caution: If you are harboring secret sin in your own life, the demons and fallen angels could turn on you, if you attempt warfare from this uncovered position. All of the prophetic gifts can be used to search out and strip this enemy, and set the people free. We must also pray that laws that need reformation are changed. Such reformation will expose deception that cannot remain hidden. For example, gambling should not be legal, nor should prostitution, or euthanasia.

The Jebusites

The tribe of Benjamin did not drive the Jebusites from Jerusalem, as God had commanded them (see Judges 1:21). The Jebusites made Mount Zion in Jerusalem their stronghold. King David captured the city of Jerusalem from the Jebusites (2 Sam 5: 6-10), and made it his capital. Mount Zion, Mount Moriah and the Mount of Olives are names of some of the most respected places in the scriptures:

- The offering of bread and wine by Melchizedek, king and high priest of Salem, took place in Jerusalem (Genesis 14:18). This was a foreshadowing of the Passover meal, when the Lord Jesus shared unleavened bread and wine with His disciples before He gave His life as a "ransom for many." The Lord Jesus took bread and said, *"This is My body, broken for you."* He took wine and said, *"This is My blood, shed for you."* It was the prefiguring of the church's communion service.
- The sacrifice of Abraham of his son Isaac took place on Mount Moriah (Genesis 22).
- The crucifixion of the Lord Jesus took place outside the walls of Jerusalem.

This is holy ground. The Jebusites had owned one prime piece of real estate, up to the time of King David. **The name "Jebusite" means treading or trodden down**. It comes from the root word to tread down with the feet, to pollute or to defile. As we now understand the meaning of their name, and we connect this with their ownership of Jerusalem, we get a picture of the activity of the "Jebusite spirit."

The Jebusite Spirit Seeks to Trod Down the Jewish People

This evil spirit is a vile, malevolent spirit, sent to tread down, pollute and profane the holy things of God. The Jewish people are God's holy people, who have an end-time purpose, before the return of the Lord Jesus. This spirit has tried to annihilate them several times throughout their history. It knows that the latter glory of the church will be greater than its former glory, when the Jewish and the gentile believers worship and minister together as one new man in the Messiah. This spirit works through instigating nations, which descended from Ishmael, Abraham's son of the flesh, to destroy God's people, the Jewish nation.

The Jebusite Spirit Seeks to Trod Down the Christian People

We have seen in recent times, in the violence seen in Islamic nations, in India and elsewhere, where angry mobs attacked and killed Christians and burnt down theie churches. This is how an entire nation under the influence of the Jebusite spirit works to "trod down" the Christian people living in such anti-Christian nations.

Intercession

The Jebusite spirit is a principality and power of darkness that rules over all the nations that hate Israel and the church. This spirit would like to see Israel and the Christian church annihilated.. God calls His church to enter into the warfare of this age, against the spirit of this age. The prophet Daniel's intercession opened heaven for angelic assistance to defeat the prince of Persia (Daniel 10:13). This is an example for the church, which teaches us how to join heaven in worship, warfare and intercession, defeat the enemies of God and the church. The church does not war against flesh and blood, but against these forces of dark, demonic angels that control entire nations.

The church must also pray for Israel's protection and salvation, as she prays for her persecutors. Israel will face great pressure on her very existence as a nation, once more. Apart from international pressure, and internal political pressure, there will be pressure through wars. Only God and His angels can protect her. Christians must pray for Israel, as she has a God-given role to play in God's end time purposes.

"*The angel of the Lord encamps around those who fear Him, And rescues them.* (Psalm 34:7)."

Chapter 9

Putting it All Together

This chapter is about making the teachings in this book practical. Heaven is calling us to join them in worship, warfare and intercession.

We pray, Father, in Jesus' name, send Your angels to strengthen Your church. Help them to partner with You in the restoration and renewal of the earth. Awaken Your bride to the worship, warfare and intercession of heaven. Give us the grace to say, along with our Lord Jesus, "*Father, the hour has come for You to send You Son to take up His throne.*" *(Isa. 9:6-7).*

Growing passionate for the Lord's return to the earth

We must grow more passionate for the return of the Lord Jesus. When we understand how the church on earth can join with heaven to hasten His return, oh, how our hearts will burn within us for our Bridegroom to come back. He is worth it all. Our present suffering is nothing, compared to the glory revealed in us and on the newly earth. No eye has seen, no ear has heard, what awaits those who wait for Him. "*And nothing unclean, and no one who practices abomination and lying, shall ever come into it, but only those whose names are written in the Lamb's book of life.*" (Rev. 21:27).

Preparing the Church for a Different Service of Worship, Warfare and Intercession

Let's go back to the seven aspects of worship in heaven, that do not need to be sung in this order listed below. They are woven together, or sung separately. You can do this as the Spirit leads.

1. **Chorus of the Cherubim**: The chorus of the four living creatures, the Cherubim: "Holy Holy, Holy, Lord God Almighty."
2. **Chorus of the Elders** around the throne of God, lifting up prayers of intercession.
3. **Chorus of the Bride** in heaven: The redeemed sea of humanity in heaven are declaring the beauty of the Lord. These are the songs of the Spirit and Bride, calling the bride on earth to join her in heaven as one voice. There will be one bride on earth and in heaven, beckoning the Lamb on the throne to take the scroll and break the seals.
4. **Chorus of the Angels sing** The New Song of heaven: "Worthy is the Lamb who was Slain."
5. **Chorus of the tribes and nations**: "To Him who sits on the throne and to the Lamb, be praise and honor and glory and power, forever and ever!"
6. **Chorus of creation**: The Victorious Song of the Overcomers. "Salvation belongs to our God, who sits upon the throne, and to the Lamb."
7. **The Great Thanksgiving Song:** Halleluia, Halleluia, for the Lord Omnipotent Reigns!

I think the simplest way to pattern our worship, warfare and intercession based on heaven, is to begin with Jill's CD – *Sounds of Heaven*. This book and the CD used together is a very practical approach. We can begin the prayer meeting or church service with an introduction, if you like. Sing some of the songs that everyone is familiar with, so the worship leader has their attention and the congregation's spirit is moving into the realm of worship. Then look at the lyrics of the first song: "Song of the Cherubim." You will need to have the CD *(Sounds of Heaven)* to hear the song for yourself.

Jill and I have created a beautiful Power Point Presentation, which contains the lyrics to all of the Sounds of Heaven songs in one file. This will be available for free download from Jill's website: www.coffeetalkswithmessiah.com on the Music Page.

The lyric and chord sheets for each song will also be available on Jill's website as Word documents. This will help worship leaders who desire to play these songs.

Large groups/churches

Jill has given me permission to publish Song of the Cherubim here, as an example of how we can write songs with multiple choruses.

Song of the Cherubim by Jill Shannon, Copyright © 2011

CHORUS:

F#m D A E(G#)
Holy, Holy, Holy, Adonai Almighty
F#m D A E(G#)
Who was and is and is to come, Glory to the Righteous One

Harmony 1
F#m D A E(G#)
Father of lights, You shower Your mercy, Father of lights, Your pour out love
F#m D A E(G#)
Father of lights, Your mercies are new every morning, every morning

Harmony 2, added later in the song:
F#m D A E(G#)
I will sing to my king, I will love no other one
F#m D A E(G#)
All my heart, all my soul, all my strength to Him alone

Verse 1

F#m   E(G#) A   D   F#m   E(G#)   A   D
I saw an open heaven,   I saw the King upon His throne
F#m   E(G#) A   D Bm7 A(C#) Esus
I heard the living creatures crying Holy,   Holy,   Holy

Verse 2 (same chords as verse 1)

I saw Seven Spirits of God like fire, I saw them sent into all the earth
I saw the elders bowing low, and casting down their crowns

Bridge 1
        D   F#m   E   A(C#)   D   F#m   E
Like a jasper and a sardius stone, The Shining One upon His throne
  D   F#m   E   A(C#)   D
With the sounds of wings and Seraphim of fire, In the thunderings
  F#m   E
of His desire
        D   F#m   E   A(C#)   D
O Lamb of God, You are worthy, O the beauty of Your heart
 F#m   E
of mercy
                D   F#m   E   A(C#)   D
Transparent Jewel, You reign alone, Ancient of Days, let Your
  F#m   E
Kingdom come!

CHORUS

Verse 3 (same chords as verse 1)
Give thanks to the Lord, for He is good,
Give thanks to the Lord for His mercies endure
Give thanks to the Lord for His mercies are new,
They're new for us each morning

Bridge 2 (same chords as Bridge 1)

Here at your feet, we cast our crowns, With all of our strength, we cast them down
My soul must sing to the Holy One, My soul must sing to the Only One

CHORUS and singing in the Spirit

How to use the multiple choruses
I have added my interpretation of the seven aspects of worship alongside the different parts of the lyrics.

Begin with an opening interlude of instrumental music (optional: with the sound effects of lightning and thunder, the sound of the flow of "many waters," as it is in heaven). Next, a solo voice follows, like that of Jill's or even a male voice, singing:

I saw an open heaven
I saw the King upon His throne
I heard the living creatures crying… (The Song of the Cherubim now begins and continues as the solo is being sung, but a bit faint so that the solo can be heard)

I saw Seven Spirits of God like fire
The eyes of the Lord into all the earth
I saw the elders bowing down, casting down their crowns… (The Song of the Elders now begins, the Song of the Cherubim continues in the background, but can still be heard, as the solo voice continues)

Like a jasper and a sardius stone
The Shining One upon His throne
With the sounds of wings and Seraphim of fire
In the thunderings of His desire… (The Song of the Angels follows, sung by the myriads and myriads of angels in heaven – on earth the entire church can sing this part – however, the song of the Cherubim and The Song of the Elders need to be interwoven into the pattern)

The Chorus of the Cherubim, the Four Living Creatures. In a large choir that can do multiple choruses, one group can sing this part, right through the entire song. These same people can also sing the other parts of the song, creatively adapting the tunes to create beautiful harmonies.	Holy, Holy, Holy, Lord God the Almighty Who was and is and is to come Glory to the Righteous One

The Chorus of the Elders. This is the depth of prayer that should resound from the hearts of the church elders – calling out to the Father, seated on the Mercy seat. This sets the place for intercession for the nations. Let the Holy Spirit lead the intercession for the nations. You can declare your intercessions or sing them to the Lord in your own melodies or in tongues.	Father of lights, You shower Your mercy Father of lights, Your pour out Your love Father of lights, Your mercies are new every morning Here at your feet, we cast our crowns, With all of our strength, we cast them down My soul must sing to the Holy One, My soul must sing to the Only One

The Chorus of the Angels. Here, we are just declaring the beauty and the worthiness of God, and merciful heart of our Lamb of God.	O Lamb of God, You are worthy O the beauty of Your heart of mercy Transparent Jewel, You reign alone Ancient of Days, let Your Kingdom come!

The Chorus of the Tribes and Tongues and nations. Like the elders, we all know that we do not deserve to wear the crown the Lord gives us. We cast it down at His feet, and let your soul sing to the One you love. Just let Him know that all the glory belongs to Him.	With all of our strength, we cast them down My soul must sing to the Holy One My soul must sing to the Only One

The Chorus of the Bride. The song of the Bride is what heaven is waiting to hear –The church expresses her passion for the return of the Lord Jesus, her bridegroom.	I will sing to my king, I will love no other one All my heart, all my soul, all my strength to Him alone

Warfare and intercession. This is the ONE SOUND of entire church should move into intercession, in spoken prayers, or sung intercessions, or declaring Scriptures aloud, and judging the powers of darkness. In the Glory, it is safe to do warfare, under the cover of the Lord.	

This last part: The great Thanksgiving, after the warfare and intercession

Heaven and earth resound in one voice, to the Lamb of God who has the victory in the battle. All groups sing together. You can add many other words of thanksgiving to these words below. They are just a sample.

Give thanks to the Lord, for He is good, Give thanks to the Lord for His mercies endure Give thanks to the Lord for His mercies are new, They're new for us each morning

Small groups / Churches

If you don't have enough singers to handle multiple choruses, then simply sing the song through and allow the Spirit of the Lord to move the group (through the leader/pastor) into intercession. Whether you sing the songs from Jill's CD or you compose your own, or you sing the more familiar worship songs, they should express the present activity in heaven, as described in the book of Revelation. Being one with God's will is to do what He is doing in heaven and on earth, today.

Intercession Should Flow from the Heart of Worship.

Here is a simple example of how worship and intercession can flow in and out, as well as together.

1. After the entire congregation has reached a depth and beauty of worship, the elders or prayer leaders should then initiate spoken intercession.
2. The music ministry should continue playing background instrumentation. In some cases, they can continue singing, but in a lower volume, so that the prayers of intercession can be heard.
3. The music ministry and the elder/leaders should work out singing between prayers of intercession and the songs of worship. For example, one of the leader/elder begins to lead with an intercessory prayer. The rest of the church joins in this prayer. Among the singers on the worship team, there should be one or two who might respond to the spoken prayers with a "sung prayer." This would be a spontaneous melody and prayer from one's heart which completely flows from the spoken themes. This is called "responsive singing" or "antiphonal singing." After a few more similar intercessory prayers, the music team can come back in with a song/chorus

and the volume can be increased again to the normal levels. This pattern will usually have two or three cycles during a service, i.e., worship, prayer and spontaneous singing, worship songs, prayer and spontaneous singing, singing in the Spirit as well (tongues).

The intercession can be lead in a variety of ways:

1. The elder/leader can speak about an incident, news article or something the Lord is speaking about that needs urgent intercession. Large groups can put up the intercession on a projector overhead.
2. He/she can lead the intercession and the entire church can be called to pray about that intention
3. The prophetic word can also be used to focus the intercession.
4. The elders/leaders can prepare the intercession prior to the date of the service, based on a current situation, i.e., the Japanese earthquake

Other Deeper Ways of Intercession

The Lord may give you a nation to claim for the gospel. Adopt that nation for a season of prayer. Learn about that nation, its people, its main religion, its history, its cities, its founding fathers and their philosophy of life that is imprinted into the life of the city, through its architecture, the design of the city and way of life. George Otis Jr. coined a phrase "spiritual mapping," which is all about learning about the nation while keeping your senses open to the Holy Spirit to help you read between the lines of nation's story. There are some excellent books in the market on "spiritual mapping."

Cry out to God through the time of worship and intercession, for this nation and its people. If you can identify the "Pharaohs" (the controlling spirit of darkness over the nation), then like Moses, command it, "let the people go." [59] In the spirit, tear down manmade altars to this controlling spirit. Pray for the people of God in this

[59] Just a word of caution here: When applying Paul's teaching in Ephesians 6:10, concerning principalities and powers of darkness in high places, it is wise to deal with them as a church, the body of Christ. Along with your pastor, there is a matched spiritual authority to deal with these spirits that control nation/s. I have

nation, the church. Cry out to the Father, for this is the bride of His Son in that nation. Pray for her protection, and pray for her to stand firm in God. Pray that she will intercede with heaven, for heaven is waiting to hear her voice and come to her assistance. The Lord Jesus will bring His righteous judgments upon the powers of evil and upon the enemies of God. Ask the Holy Spirit to bring forth restoration and renewal in that nation, in its cities, in the church. Try to "keep in touch" with what is happening with that nation, its cities, its people, the church.

You can keep your research notes and compare them with the others in your group/church. When you gather to worship (remember to pattern it as it is in heaven), you can bring your thanksgiving to God, in the great Hallelujah. Each time the church or group enters the heavenly pattern of worship, warfare and intercession for that nation, the kingdoms of the earth are yielding to the power of God. He is at work, stripping the earth of evil, restoring, and renewing it as the kingdom of our Lord. Keep at this pattern in worship and intercession for that nation, until the Holy Spirit says, *"Now, look for another nation."*

Intercession Guided by the Council of God in Heaven

"But who has stood in the council of the Lord, that he should see and hear His word? Who has given heed to His word and listened?" (Jer.23:18).

In the realm of the spirit, prophetic intercessors can come before the council of the Lord in heaven, and can inquire about an assignment for intercession. This is another way that the church on earth can join with the intercession in heaven, and move with one accord. It is a very powerful way to intercede, as the council of the Lord, which consists of living saints in the presence of God, have the mind of Christ. They are able to guide the church on earth and draw her to be one with the intercession in heaven. If your group is so blessed as to have a prophetic intercessor who is taken before the Council in

---

seen back lash when individuals try to take on such high level spiritual powers of darkness, and it can be a very hard blow.

Heaven, then it is important to follow their instructions, which they received from the council of the Lord. Discernment of spirits is an important gift of the Holy Spirit needed here. So also is walking in humility.

Intercession for the End of the Ages

As we move towards the end of this age, and prepare for the return of the Lord to the earth, the church must move with God in what He is doing. The word of God is our sure guide for intercession. We will need to interpret the signs of the age, and to come before the Lord to understand what He is doing in the earth, then seek His mind to join Him in His holy assignment.

As an example, let us suppose that the Lord shows us that He has released the red horse rider into the world. How can the church join with the Lord in this act of releasing of the red horse rider? In the word of God, the angel riding the red horse is instructed go and to take away peace from a nation. As we read the daily news, we get an emerging picture of the pain and suffering in many nations around the world. Many nations are experiencing devastating natural disasters. Haiti, Chile, Christchurch in New Zealand, and more recently China and Japan had major earthquakes, and Japan also had a tsunami, where the dead and missing have risen to staggering numbers. People in these places have lost their belongings, property, and loved ones. Has the rider of the red horse visited these nations taking away their peace? How do we pray in the midst of such sorrow and suffering, pain and hardship?

The rider of the red horse comes with God's partial judgment. We need the mind of Christ to pray into these situations. God knows what He is doing; man cannot fathom the mind of God. Our Lord Jesus understands the situation, as He Himself was a man of sorrow

We can pray that through all this pain and sadness, that people will turn to the Lord Jesus and commit their lives to Him. Many will question the love of God at these moments. Many others question the very existence of God, saying, "Is there a God who cares?" God has seen their sorrow in all these events in these nations, since the day His Son, our Lord Jesus Christ, hung upon the cross. When He died for their sins, He has carried all of this grief, reaching out to all men, despite their rejection of His love. The Lord Jesus walks among them and is with them even now in their sorrow. He moves among them and feels their pain, with the hope that some will recognize Him and respond to His redeeming love. Our intercession is that even in this time of judgment, the nation will respond to His saving grace.

As part of the discernment process as we pray for the return of the Lord, it is good to keep a journal of your intercessory events and how the nation responded. This will mean watching for signs of change. Surprisingly, this may come in a variety of ways, such as news events on TV, newspapers, magazines, radio, or books. As a "prayer watchman," you are constantly looking out for that coming change.

For every change, have a "praise" party with heaven! (see Luke 15:10). Keep up the good work as you pattern your worship, warfare and intercession as it is in heaven, before the throne of God.

Let us take another example: the rider on the white horse goes about his task of proclaiming the Word of God. Ask the Father, "Father which nation(s) needs to hear the gospel, the saving word of God?" Or, pray for nations currently in the grips of chaos. "Lord, in the midst of all the chaos in Japan, may Your healing love go out to touch, heal, and draw the broken-hearted to the love and healing of Your Son, our Lord Jesus Christ. Lord, show us how to pray for a nation so devastated by the earthquake and tsunami, that the saving grace of the gospel would bring a ray of hope to the terrible pain in their hearts and spirits. Lord, we pray for those in authority, for wisdom, courage and strength in their leadership. May Your Holy Spirit give them the wisdom to do what is right, and lead them to Your saving word. AMEN."

Appendix

THE AMIDAH

(Means "Standing" and refers to the fact that the prayer is said standing up)

1. The God of history:

Blessed are You, O Lord our God and God of our fathers, the God of Abraham, the God of Isaac and the God of Jacob, the great, mighty and revered God, the Most High God who bestows lovingkindnesses, the creator of all things, who remembers the good deeds of the patriarchs and in love will bring a redeemer to their children's children for his name's sake. O king, helper, savior and shield. Blessed are You, O Lord, the shield of Abraham.

2. The God of nature:

You, O Lord, are mighty forever, You revive the dead, You have the power to save. [From the end of Sukkot until the eve of Passover, insert: You cause the wind to blow and the rain to fall.] You sustain the living with lovingkindness, You revive the dead with great mercy, You support the falling, heal the sick, set free the bound and keep faith with those who sleep in the dust. Who is like You, O doer of mighty acts? Who resembles You, a king who puts to death and restores to life, and causes salvation to flourish? And You are certain to revive the dead. Blessed are You, O Lord, who revives the dead.

3. The Sanctification of God's Name:
[Reader] We will sanctify your name in this world just as it is sanctified in the highest heavens, as it is written by your prophet: "And they call out to one another and say:
[Cong.] 'Holy, holy, holy is the Lord of hosts; the whole earth is full of his glory.'" [Isa. 6:3]
[Reader] Those facing them praise God saying:
[Cong.] "Blessed be the Presence of the Lord in his place." [Ezek. 3:12]
[Reader] And in your Holy Words it is written, saying,
[Cong.] "The Lord reigns forever, your God, O Zion, throughout all generations. Hallelujah." [Ps. 146:10]
[Reader] Throughout all generations we will declare your greatness, and to all eternity we will proclaim your holiness. Your praise, O our God, shall never depart from our mouth, for You are a great and holy God and King. Blessed are You, O Lord, the holy God. You are holy, and your name is holy, and holy beings praise You daily. (Selah.) Blessed are You, O Lord, the holy God.

4. Prayer for understanding:
You favor men with knowledge, and teach mortals understanding.
O favor us with the knowledge,
the understanding and the insight that come from You.
Blessed are You, O Lord, the gracious giver of knowledge.

5. Prayer for repentance:
Bring us back, O our father, to your Instruction;
draw us near, O our King, to your service;
and cause us to return to You in perfect repentance.
Blessed are You, O Lord, who delights in repentance.

6. Prayer for forgiveness:
Forgive us, O our Father, for we have sinned;
pardon us, O our King, for we have transgressed; for You pardon and forgive.
Blessed are You, O Lord, who is merciful and always ready to forgive.

7. Prayer for deliverance:
Look upon our affliction and plead our cause,
and redeem us speedily for your name's sake,
for You are a mighty redeemer.
Blessed are You, O Lord, the redeemer of Israel.

8. Prayer for healing:
Heal us, O Lord, and we will be healed;
save us and we will be saved, for You are our praise.
O grant a perfect healing to all our ailments,
for You, almighty King, are a faithful and merciful healer.
Blessed are You, O Lord, the healer of the sick of his people Israel.

9. Prayer for deliverance from want:
Bless this year for us, O Lord our God,
together with all the varieties of its produce, for our welfare.
Bestow ([from the 15th of Nissan insert:] dew and rain for) a blessing upon the
face of the earth. O satisfy us with your goodness, and bless our year like the best of years.
Blessed are You, O Lord, who blesses the years.

10. Prayer for the gathering of exiles:
Sound the great shofar for our freedom,
raise the ensign to gather our exiles,
and gather us from the four corners of the earth.
Blessed are You, O Lord, who gathers the dispersed of his people Israel.

11. Prayer for the righteous reign of God:
Restore our judges as in former times, and our counselors as at the beginning; and remove from us sorrow and sighing. Reign over us, You alone, O Lord, with loving kindness and compassion, and clear us in judgment. Blessed are You, O Lord, the King who loves righteousness and justice.

12. Prayer for those who are against God:
Let there be no hope for slanderers,
and let all wickedness perish in an instant.
May all your enemies quickly be cut down, and may You soon in our day uproot, crush, cast down and humble the dominion of arrogance. Blessed are You, O Lord, who smashes enemies and humbles the arrogant.

13. Prayer for those waiting for God
May your compassion be stirred, O Lord our God, towards the righteous, the pious, the elders of your people the house of Israel, the remnant of their scholars, towards proselytes, and towards us also. Grant a good reward to all who truly trust in your name. Set our lot with them forever so that we may never be put to shame, for we have put our trust in You. Blessed are You, O Lord, the support and stay of the righteous.

14. Prayer for the rebuilding of Jerusalem:
Return in mercy to Jerusalem your city, and dwell in it as You have promised.
Rebuild it soon in our day as an eternal structure,
and quickly set up in it the throne of David.
Blessed are You, O Lord, who rebuilds Jerusalem.

15. Prayer for the coming Messianic King:
Speedily cause the offspring of your servant David to flourish,
and let Him be exalted by your saving power,
for we wait all day long for your salvation.
Blessed are You, O Lord, who causes salvation to flourish.

16. Prayer for the answering of prayer:
Hear our voice, O Lord our God; spare us and have pity on us.
Accept our prayer in mercy and with favor, for You are a God who hears prayers and supplications.
O our King, do not turn us away from your presence empty-handed,
for You hear the prayers of your people Israel with compassion.
Blessed are You, O Lord, who hears prayer.

17. Prayer for the Jerusalem Temple Service:
Be pleased, O Lord our God, with your people Israel and with their prayers.
Restore the service to the inner sanctuary of your Temple,
and receive in love and with favor both the fire-offerings of Israel and their prayers.
May the worship of your people Israel always be acceptable to You.
And let our eyes behold your return in mercy to Zion.
Blessed are You, O Lord, who restores his divine presence to Zion.

18. Thanksgiving prayer for the mercies of God:
We give thanks to You that You are the Lord our God
and the God of our fathers forever and ever.
Through every generation You have been the rock of our lives, the shield of our salvation. We will give You thanks and declare your praise for our lives that are committed into your hands, for our souls that are entrusted to You, for your miracles that are daily with us, and for your wonders and your benefits that are with us at all times, evening, morning and noon.
O beneficent one, your mercies never fail; O merciful one, your loving kindnesses never cease. We have always put our hope in You. For all these acts may your name be blessed and exalted continually, O our King, forever and ever. Let every living thing give thanks to You and praise your name in truth, O God, our salvation and our help. (Selah.)
Blessed are You, O Lord, whose Name is the Beneficent One, and to whom it is fitting to give thanks.

19. Prayer for peace:
Grant peace, welfare, blessing, grace, loving kindness and mercy to us and to all Israel your people. Bless us, O our Father, one and all, with the light of your countenance; for by the light of your countenance You have given us, O Lord our God, a Torah of life, loving kindness and salvation, blessing, mercy, life and peace.
May it please You to bless your people Israel at all times and in every hour with your peace.
Blessed are You, O Lord, who blesses his people Israel with peace.

The Kaddish

May the great Name of God be exalted and sanctified, throughout the world, which He has created according to his will. May his Kingship be established in your lifetime and in your days, and in the lifetime of the entire household of Israel, swiftly and in the near future; and say, Amen.

May his great name be blessed, forever and ever. Blessed, praised, glorified, exalted, extolled, honored elevated and lauded be the Name of the holy one, Blessed is He- above and beyond any blessings and hymns, Praises and consolations which are uttered in the world; and say Amen. May there be abundant peace from Heaven, and life, upon us and upon all Israel; and say, Amen.

He who makes peace in his high holy places, may He bring peace upon us, and upon all Israel; Amen.

Some web URLs for References on Revelation and Prophetic Literature

Articles, etc. Related to the Book of Revelation (List of web links)
http://www.torreys.org/bible/biblia02.html#revelation
Jewish Virtual Library – A division of the American-Israeli Cooperative Enterprise
http://www.jewishvirtuallibrary.org/index.html
The Cities of Revelation
http://www.luthersem.edu/ckoester/Revelation/main.htm
Crossroads in Prophetic Interpretation - Historicism Versus Futurism, by Gerhard F. Hasel. http://www.e-historicist.com/GFHasel/gfhcrossroads1.html
Historicism, Futurism, Preterism, by Lawrence R. Kellie. 3p.
http://www.e-historicist.com/LawrenceKellie/histfutpret.html
Inventory of Symbols Used by John in the Book of Revelation
http://www.bibleonly.org/proph/rev/symbols.html
Links to Revelation, Apocalyptic and Millennial Websites and Materials
http://clawww.lmu.edu/faculty/fjust/Apocalyptic_Links.htm

Links to Revelation, Apocalyptic and Millennial Websites and Materials
http://www.clawww.lmu.edu/faculty/fjust/Apocalyptic_Links.htm
Modern History Sourcebook: Rev. Charles Davy: The Earthquake at Lisbon, 1755
http://www.fordham.edu/halsall/mod/1755lisbonquake.html
New Testament Gateway: Book of Revelation
http://www.ntgateway.com/rev.htm
Catholic Encyclopedia: Apocalypse
http://www.newadvent.org/cathen/01594b.htm
The Catholic Origins of Futurism and Preterism
http://www.biblelight.net/antichrist.htm
The Catholic Origins of Futurism and Preterism
http://www.aloha.net/~mikesch/antichrist.htm
The Prophecies of Daniel and Revelation, by Uriah Smith
http://ourworld.compuserve.com/homepages/clt4/drtoc.htm
Revelation
http://pacificcoast.net/~muck/rev.html
Revelation Resources (Resources for the Study of the Book of Revelation)
http://www.book-of-revelation.com/
Revelation Resources: Resources for the Study of the Book of Revelation
http://www.teologi.dk/Revelation/
A Summary of the Symbols Used in Revelation
http://www.bibleonly.org/proph/rev/symbols2.html
Who Are the Jews? (Jewish motifs in the Book of Revelation)
http://www.bibleonly.org/proph/rev/jews.html
Who Should You Believe? The Four Major Schools of Prophetic Interpretation. (Author is an SDA)
http://www.bibleonly.org/proph/schools.html

About Robert Misst

Robert (Bob) Misst was baptized in the Holy Spirit in 1974. Prior to his baptism of the Holy Spirit, Bob drifted from the Christian faith while at university when he read the books of Bertrand Russell, the famous British atheist, philosopher, mathematician, scientist and prominent writer, who influenced his thinking and young mind to choose atheism over Christ in his life.

When he met the Lord Jesus Christ in 1974, he struggled hard to surrender his complete life to Christ as he was working on his professional career as an accountant. In the mean time, he served on various church renewal programs. In 1981, he helped establish a Covenant Christian Community, a branch of the Sword of the Spirit, an ecumenical International Covenant Christian Community, in Mumbai, India). In 1987, a major change took place in his ministry. God called him into prophetic intercession for the nations. He and his team lead night vigils of worship and intercession for the city, the church and the nations. This drew crowds of over 1500 people spending a whole night of worship and intercession in the heart of Mumbai city. He and his team travelled to many parts of India teaching prophetic intercession to the body of Christ. Gradually, the fire of intercessory prayer spread across the nation of India. Between 1989 and to this day, Bob has travelled with intercessory prayer teams, which include internationally acclaimed bible teachers like Derek Prince, Lance Lambert, Johannes Facius, and others to many nations across the globe. In 2002, the Lord called Bob to move to New Zealand. He and his family currently are New Zealand citizens, and live in Auckland. In 2005 with a team of 11 others, he travelled the length of New Zealand, praying for the nation to open up to the Lord. Bob works with intercessors and prophets in New Zealand. He also works with the leadership of the National Days of Prayer network.

This is Bob's first book. In this project, he has collaborated with internationally known songwriter, musician, author and bible teacher, from Philadelphia, USA, Jill Shannon who has produced the awesome worship CD, "Sounds of Heaven," which is based on Bob's teaching on worship as in heaven. Jill has also written the Foreword to the book and a chapter on the practical aspects of Christian song writing, encouraging others to follow the pattern of worship in heaven. She has most graciously spent hours editing the book in the midst of her own busy teaching and preaching ministry.

Bob's contact details for ministry purposes (Seminars on Worship, Warfare and Intercession for the return of the Lord Jesus).

e mail address is: ramisst@yahoo.com.
Telephone Number. +64 9 255 5138 (Auckland, New Zealand).
Address: 29 Secoia Crescent, Mangere, Manukau, Auckland 2022, New Zealand.

Copies of this book are available at your local Christian book-seller or asking them to order the book from Xulon Press, or from Jill Shannon's website: www.coffeetalkswithmessiah.com

About Jill Shannon

Jill Shannon is a Messianic Jewish Bible teacher, author and singer/songwriter. Growing up in a Jewish home, she accepted the Lord in 1973. In the 1980's, Jill and her husband immigrated to Israel, learned Hebrew and gave birth to three children. During these years in Israel, she endured hardship and received vital lessons, shared in her first book, "*Coffee Talks with Messiah: When Intimacy Meets Revelation.*"

Jill's second book is "*A Prophetic Calendar: The Feasts of Israel,*" (Destiny Image Publishers, 2009). Her third book is *"The Seduction of Christianity: Overcoming the Lukewarm Spirit of the Church,"* (Destiny Image, 2010). Her books are available from her website and in bookstores everywhere.

Her latest book is called, *"If I Forget Jerusalem: Israel's True Story and Eternal Destiny."* It is a comprehensive prophetic and teaching course on Israel's history, future destiny, and the Lord's heart and Kingdom purposes for His Jewish people. It should be out by the end of 2011. She has also completed filming a 12-part, high-definition DVD teaching series based on this new book on Israel, which will be out during 2011.

Check Jill's website for details. www.coffeetalkswithmessiah.com

Jill currently speaks and writes about experiencing God's glory, holy living and intimate friendship with the Lord, the biblical Feasts, Israel and the Church. She is also a worshiper/songwriter, with six worship projects to date: *"A Part of Me," "Beckon Me," "Remember Me," "I AM the Broken Piece,"* and her "soaking" CD, *"Song of the Lamb."* Jill is currently producing her newest worship CD, *"Sounds of Heaven,"* which will also be released mid-2011. You can listen to clips of her music on her website.

Jill presently resides outside of Philadelphia, Pennsylvania with her husband, and will also be residing in Israel in the near future. Jill has a married son and daughter-in-law, two daughters in ministry school, three grandchildren, and one grandson in heaven. To order or learn more about her books and listen to clips from her worship CD's, and to listen to free teachings, go to her website, www.coffeetalkswithmessiah.com

CPSIA information can be obtained at www.ICGtesting.com
Printed in the USA
LVOW071659201011
251254LV00003BA/2/P